THEORIES OF IDEOLOGY AND IDEOLOGY OF THEORIES

POZNAN STUDIES
IN THE PHILOSOPHY CF THE SCIENCES AND THE HUMANITIES

VOLUME 9

The principal task of the book series "Poznań Studies in the Philosophy of the Sciences and the Humanities" is to promote the development of the philosophy which would both remain within the Marxist tradition of great philosophical ideas and respect the manner of philosophical thinking introduced by the 20th century positivism. Our aim is to contribute to practicing philosophy as deep as Marxism and as rationally justified as positivism is.

Address of the editor:
Leszek Nowak
Department of Philosophy   Book series "Poznań Studies in the
UAM   Philosophy of the Sciences and the
Szamarzewskiego 89C   Humanities" is partly sponsored by
60-568 Poznań, Poland   Adam Mickiewicz University

THEORIES OF IDEOLOGY
AND
IDEOLOGY OF THEORIES

Edited by

Piotr Buczkowski and Andrzej Klawiter

AMSTERDAM 1986

CIP-GEGEVENS KONINKLIJKE BIBLIOTHEEK, DEN HAAG

Theories

Theories of ideology and ideology of theories. — Amsterdam
: Rodopi. — (Poznań studies in the philosophy of the
sciences and the humanities ; vol. 9)
ISBN 90-6203-837-9
SISO 134 UDC 165
Trefw.: ideologie.

© Editions Rodopi B.V., Amsterdam 1986
Printed in The Netherlands

TABLE OF CONTENTS

I. THEORIES OF IDEOLOGY

Piotr Buczkowski/Poznań

THE LEVELS OF CONSCIOUSNESS

Some remarks on sociology of knowledge

1. Introduction

Sociology of knowledge is a subbranch of social studies devoted to problems of social consciousness. I shall tackle the difficulties of determining an empirical status of social consciousness in the second part of my paper, limiting myself at present to the distinction of three planes of research on these problems. Thus one can study the mechanisms of generating beliefs which constitute social consciousness, the mechanisms of their proliferation or social transmission (how are they coded, sent, accepted by individuals), and their social functioning (i.e. relations between states of consciousness and practical activities). Theoretically speaking, a sociologist of knowledge has to investigate these three planes of social consciousness, either attempting to encompass all three or one of the above-mentioned three planes.

Generally speaking, most sociologists of knowledge focus on one aspect, singled out for analytical purposes. R.K. Merton (1982, 477) pointed out a difference between European and American sociology of knowledge, stating that sociology of knowledge seems to be an European speciality, while sociology of mass communication remains an American one. From the point of my distinction the difference lies in different planes of analysis taken separately. The so-called sociology of knowledge in its characteristically European development focused on social problems of knowledge production, especially on the influence of social structure upon theoretical constructs. Differences of concepts of social structure in particular cases are secondary. The Marxist tradition evokes class structure, i.e. the determination of consciousness by material conditions of life common to huge social groups — namely classes (which — in the ultimate resort of such analysis — makes one look for historically mutable relations of production).

For instance A. Schaff (1970, 175-176) writes: "Whoever claims that base influences changes in superstructure acknowledges social conditions of consciousness and its changes. This general thesis is made more concrete as applied to social groups and individuals within them with the aid of a theory of a class nature of consciousness and cognition. If one acknowledges social conditioning of consciousness because of its dependence on base, on social being, then one also acknowledges the influence of the relations of production — which are the core element of a base — upon consciousness. The relations of production, especially property relations, decide class divisions. Classes represent definite interests which influence also human cognitive attitudes. Since interests can differ, and they can also be contradictory, their influence upon cognition results in different products of cognitive activities." Let us note that according to Schaff deformations of consciousness result from reflections   of social being (relations of production) and from "projections" of various class interests. Both parts of this explanation can be questioned. First, in what sense does being reflect itself in a scientist's mind? Scientists usually study some aspects of reality (including economists with a proud exception of Ricardo) and neglect to notice economic base or relations of production organization. They can be aware of property relations, but these are just an aspect of the relations of production (Buczkowski, 1982, 160 and ff.). Moreover, the property relations are perceived not in their real, economic form, but in the sphere of civil law. In this particular sense the perception of property relations remains constant in all socio-economic formations distinguished by Marx.

The second part of Schaff's explanation of the deformation of consciousness also gives rise to many doubts. In the theory of historical materialism one assumes some criteria of distinguishing social classes. As a rule one divides society into "owners" and "propertyless" direct producers. All remaining class divisions occur within subclass divisions. However, one may ask in what sense does a class interest of industrial capitalists differ from commercial or banking capitalist class interests? All these subclasses have a common major value, i.e. they strive for profit, thus defining also their basic class interest. Their interests do conflict (in that, for instance, the magnitude of industrial profit is an inverse of banking interest rates), but they are not different, nor are they oppositional. Over a long period of time, as Marx said in *The Capital* already, various types of surplus value settle down at the same level. This approach is not compatible with the classics, who wrote on spiritual authority and ownership of means of spiritual production (Marx, Engels, 1975, 50): this ownership, according to them, influences not the form of consciousness but an ability to produce certain world-outlook proposals and to disseminate them.

Another manner of thinking, originated by M. Weber with his analyses of value-judgements in science, points out definite assumptions determining social consciousness, especially the value systems which define social position and directly influence theoretical constructs. One of the descriptions of this attitude runs as follows: "The basic problem of sociology of knowledge can be reformulated as follows: what is the relation between axiological structures accepted by a given group and essential structures established by a scientist who represents it (...) what is the relation between value systems accepted by a group and systems of magnitudes assumed by its theoretical representative." (Nowak, 1974, 94 — 95). Let us notice that as far as influence of value systems is stressed (or an influence of social being in various Marxist shades of that thesis), the problem is to study falsifications of knowledge, to question epistemological status of beliefs which are produced, to consider concepts and theses in terms of truth and falsity. This fact made Kolakowski (1982) ask about epistemological meaning of the statements made by sociologists of knowledge. Arguing against K. Mannheim, Kolakowski tries to undermine the legitimacy of this manner of studying social consciousness. "The question we pose is the following one: does the influence of extra-cognitive situations upon the contents of our knowledge matter for epistemological truth-falsity one?" (ibid., 35). Mannheim's answer was, as we know, in the negative. Kolakowski reduces the problem to mutual relations between a question about genesis of definite statements and their epistemological status, i.e. their truth or falsity. While supporting Kolakowski's view that epistemological questions could be settled only if we had an ideal model of an undeformed reality, which would presuppose a privileged position of our investigator, let us nevertheless point out that a sociologist of knowledge has to study not only mechanisms of deformation of knowledge, but also its social functioning. From the latter point of view the question about truth-falsity of some statements is inferior as compared to the question if they are — to employ Kmita's formula (1982, 63 and ff.) a "regulator" of respective types of social practice. From Kmita's point of view if a primitive society accepts mythological statements encouraging to undertake magical activities within the productive practice, then this acceptance is vital — and the awareness of inefficiency of these activities (which we have, but the society in question hadn't) is of a secondary importance.

Another subbranch of sociology of knowledge — namely a sociology of science also belongs here. Mechanisms of the making of scientific knowledge are being studied within it, i.e. epistemological questions asked and mechanisms of the functioning of scientific knowledge in "scientific community" investigated. Sociologists of science also follow (varieties depend on the sociologists' interests and preferences) the generation, dissemination and functioning of scientific knowledge. What makes them different from the sociologists of knowledge is that they limit their studies to a certain group, whose social role in a macro-scale consists of producing knowledge, or to the cognitive aspects of consciousness of other groups, the ones which do not produce knowledge professionally. In my view, to be elaborated later on, a sociological analysis of knowledge should begin with the study of science's functioning — i.e. its generation, dissemination and functioning within the scientific community. However, the reconstruction of these mechanisms reveals only one of the elements of a sociological analysis of the problems of the making of social consciousness, i.e. the mechanism of the making of social knowledge.

Further problems, as I have indicated above, include the mechanisms of dissemination of knowledge which had been produced by specialists and its application in practical activities of social groups of which society is composed.

The mainstream of sociology of mass communicaton (labeled "American" by Merton) deals with the second of these problems, i.e. of the dissemination (sometimes an analysis of social functioning is also provided). Studying this problem we encounter significant phenomena, but this is not all sociology of knowledge is about. Problems of dissemination include phenomena of ideology, mass propaganda, etc., i.e. those, which refer to general concepts of social order, class relations and class domination. The relations between a message and its reception also belong here.

My own idea of a definition of the subject matter of sociology of knowledge is that this branch of sociology should study the mechanisms of the making, disseminating and functioning of knowledge which creates social consciousness of those social groups which together compose a global society. My concept of knowledge includes both substantial statements on empirical reality (their cognitive legitimacy notwithstanding) and axiological beliefs or rules of practical action.

Some assumptions should be mentioned at this juncture with respect to the empirical status of social consciousness. They will be presented at length later on, but for clarity's sake we should outline them here as well. I assume that the concept of social consciousness as a "being" which exists autonomously and independently of individual consciousness is at best a metaphor with respect to the contents of individual consciousness commonly shared (but accepted to a varying degree) by members of definite social groups. "Only human individuals are carriers of knowledge or consciousness, so there is no consciousness which would be independent of an individual and which would exist beyond or between human individuals" (Ziólkowski, 1982, 41). "Social consciousness" is not a theoretical construct of an idealized group or of an ideal subject (as

Nowak assumed with respect to class consciousness, 1984) either. It is a set of some contents shared by empirical individuals who constitute some groups or society as a whole. However, we are not dealing with a set of all beliefs held by empirical individuals (as assumed by positivistically oriented sociological research) — we deal with a subset of all contents of consciousness accepted by an individual. This subset is selected by social mechanisms of the making and disseminating of knowledge.

Having thus introduced the topic let us ask the question about the title. The concept of world outlook is probably most helpful at this point. One of the definitions of world outlook states that it is "a system composed of a set of descriptive sentences (knowledge), ethical system and the set of rules. The whole system has a following property: knowledge guarantees the existence of values distinguished in ethical system and feasibility and efficiency of actions undertaken according to the rules" (Nowak, 1974, 117). In other words, world outlook is made of knowledge (in descriptive sense), values and rules of action, which, according to this knowledge have to result in establishing the accepted values. This concept of world outlook may help us in describing the structure of individual consciousness. However, it may also be employed in order to provide a theoretical analysis of group consciousness (in spite of assumptions with respect to the status of the latter). In case of social groups I will speak of group ethos. Group consciousness or group ethos are made of properly relativized systems of knowledge, values and group action directives (or individual action directives valid within a given group). Before we analyze ethos, we should present the concept of social structure I assume in the present paper.

2. Material moment as the basic component of social structure

Adaptive interpretation of historical materialism (presented in: Buczkowski, Nowak, 1979, Buczkowski, Klawiter, Nowak, 1982) is my starting point. According to this interpretation the Marxian historical materialism, which assumes global dependencies describing economic structure of a society, can be broadened by an insertion of the category of labour division within it. This division, historically speaking, results in an appearance of some social systems, which the classics termed "material (historical) moment". These systems, which appear in various domains of divided labour, are formally analogous to each other both in inner structure and the relations between its elements. The name of "material moment" could be applied to social systems with the following properties:

(1) they have to be spheres of social activity which reflect (in their inner structure) a global structure of social life and they have to be composed

of three levels: definite tools, a set of relations between humans using them, i.e. an institutional structure which regulates these relations and consciousness (knowledge and values) of people who work within these spheres,

(2) the relations which determine peculiarities of these spheres of social life have to be isomorphic with respect to the global social relations. The system of human relations adapts itself to the level of tools, organizational (institutional) structures adapt themselves to tools and human relations and finally the consciousness within a given sphere assumes a shape which stabilizes these entities,

(3) these spheres merit the label of "material moment" since in the "ultimate resort" they are determined by the level of development of tools (if we consider them as isolated entities): each significant change of tools causes respective changes of the other structural elements,

(4) each of these spheres has a division into some social categories: into those who decide to use some tools and those, who use them.

The modification of this interpretation presented by L. Nowak (1983) indicates that the distinction of basic material moments is linked to the appearance of three antagonistic pairs: owners — direct producers, rulers — citizens, priests — believers, which occur in their respective spheres of social life: economy, politics and spiritual domain. Their interests are separate: owners try to exploit direct producers, rulers strive to control citizens' activities, while priests want their doctrines to have maximum range. These interests presuppose also some values, which can be ascribed to a collective activity of some social groups. The distinction of the three basic material moments is based on one of the meanings of the concept of the division of labour (Buczkowski, 1979). However, the same category also functions in a different meaning in the writings of the classics. In the latter case we speak of some subsystems within the larger spheres — material moments, which, although meeting the conditions 1 — 4, apply to more detailed domains of human activities. For instance, within the economic moment industry, commerce and banking emerge as autonomous areas of human economic activity. Social groups within these areas are class fractions. The fractions, as well as classes can be ascribed their proper value systems linked to their interests pursued on a social scale.

An antagonistic image of social structure is thus assumed. The basic division is into antagonistic classes which either have or have not respective material means: of production, of coercion and of indoctrination and knowledge production. Within each class a further division into fractions takes place. Activities of various categories of humans determined by their interests presupposes their acceptance of definite value systems (Buczkowski, Nowak, 1980). Each of the distinguished social groups can be ascribed its social ethos. Ethoses will differ with respect to knowledge systems, values and rules of action ordered by the former. Let us assume for the time being that differences consist of various types of social activity. Obviously, particular components of an ethos of an enterpreneur or of a worker will be different from the components of the rulers' or citizens' ethos. I do not mean they have no common features, ex. common general world image or standard types of activities undertaken in order to — let us to say — satisfy the basic biological needs, etc. Similarity of such elements, however, is from the point of my paper, less significant than the differences.

According to our discussion of consciousness of various social groups, the moment of spiritual production emerges as the most significant material moment. Within this moment knowledge which later makes various ethoses is being produced. Inner division of labour within a scientific community, its differentiation into branches or types of sciences generally corresponds to the division of society into various groups. This does not mean that everything produced within the moment of spiritual production automatically becomes a component of each particular ethos. Scientists' concepts are subjected to various selections, both inner and external ones, for instance by the existing social order, before they enter social consciousness. They are, moreover, modified in the course of direct dissemination and in the process of knowledge distribution, i.e. education, upbringing, etc. The starting point of selecting mechanisms is in most cases knowledge and value systems generated by people who directly produce them. An analysis of the functioning of science will therefore open our presentation.

3. Sociology of science as a theory of the production of social consciousness

The questions which one faces while analyzing the functioning of science refer to the generation, dissemination and functioning of scientific knowledge. Two former belong, I think, to the sociology of science proper, while the third one belongs not only there but also to the sociology of knowledge (since it studies the mechanisms of proliferation of cognitive results in society).

Within the Marxist tradition one often assumes that a theoretical consciousness is radically different from social one (i.e. the consciousness of the scientists should differ from the consciousness of rank and file members of society). In some extreme cases one juxtaposes the so-called common sense of society at large and the consciousness of the sages who are better educated and understand things more profoundly. In less restrictive interpretations social consciousness is meant to include contents which contribute to the prolongation of the social status quo. In other words, rank and file member of society has a consciousness which secures his proper performance in various social roles and also prevents him from understanding the mechanisms of society's functioning which favour some privileged groups and subsume his own interests to the interests of these groups. The interpretations of the latter kind are usually focused on the second partner of antagonistic social relations — i.e. on groups subjected to various kinds of domination. However, if one assumes that social consciousness favours contents which are functional with respect to the interests of the decision-makers with respect to some material means, then one should also demonstrate the mechanisms of dissemination and the functioning of some theoretical concepts within the scientific community. As has often been demonstrated (Kolakowski, 1978) that it is wrong to consider creative potential of scientists to be a result of the current economic situation, or, more broadly, of material conditions of social being. The very ethos of a scientist, which remains stable from the very outset of theoretical reflection, encourages to pursue other, not only pragmatic, values.

It would be, for instance, hard to deny (in spite of some attempts in the history of Marxism) that scientists actually do want to learn, understand and explain the phenomena which are of interest to them. It would also be hard to claim (in spite of the presence of such motives) that scientists want to "flatter" organizers of social order at any price, even at the cost of scientific legitimacy. As M. Polanyi once remarked (1951, 6) in opposing the thesis on direct correspondence between a need for some direction of studies and material conditions, which prevailed in the British Marxism of the 30ies: "We have to reinforce the belief that love of knowledge is the essence of science and that utility of this knowledge is not vital to us. We should demand public respect and public support for science just because it means a quest for knowledge. As scientists we have pledged loyalty to values which are much more precious than material well being and much more important." The problem is whether the results of scientific investigations correspond to their plans, and if not what "deforms" scientific theories when they are being forged. Individual views of scientists are best expressed in Weber's words on "the vocation of science" which put science above social rules as a vocation of those who work within it. Weber's aim was to free science of non-scientific values, i.e. the ones which were not linked to an explanation of a phenomenon which occupies a scientist. However, even the very system of values a scientist accepts (Buczkowski, 1976) does "deform" theoretical constructs when theory construction is attempted.

The followers of an anti-individualistic approach towards cognitive processes usually deny any influence of aims, plans and intellectual capacities of an individual upon cognitive operations. In Popper's terms, one assumes the existence of the "third world", which, in the Marxist tradition, is some content of social tradition in science (shaped by social conditions) influencing an individual in his cognitive actions. As K. Mannheim (1952, 5) put it: "Each individual is thus doubly predetermined by the fact that he or she has been brought up in a society. Ready-made made situation awaits him, and within this situation — preshaped models of thinking and behaviour". If we assume that individual thinking is univocally determined by conditions, we also have to agree that there can only be one true science in a society (considering a privileged position of intellectuals in Mannheim's approach), since a scientist has to choose the best available theory. Thus empirically visible variety of theoretical constructs would be impossible. Whoever would attempt to interpret any phenomenon, would also encounter univocal solutions, accessible to every individual who decides to start cognition. This clearly is not the case. Even Weber stressed the limitations of theoretical knowledge (1946, 138) writing that "everybody knows everything we had achieved in science will become obsolete in 10, 20, or 50 years. This is the fate of science and this is the meaning of scientific work (...) Every fulfilment in science leads towards new questions, demands to be overcome and to be turned obsolete. Everybody who wants to serve science has to acknowledge this. Scientific works can provide a lasting satisfaction because of high quality or educational value. But scientifically speaking they have to be overtaken by other works since — let us stress it again — this is our common fate and, moreover, our common aim. We cannot work without hope that other will get further than we did. Such progress is in principle infinite."

Should we assume to simplify matters that scientists work the concepts out quite independently and motivated with cognitive aims alone (i.e. to understand and explain reality), which Weber had pointed out, then social determinants of scientific creativity on the individual level would manifest themselves in value systems which the scientist had accepted. However, a value system does not predetermine a definite theoretical construct. It indicates a class of solutions, and from this class a scientist choses one according to purely cognitive criteria (explanatory power of a theory, empirical evidence, prediction, etc.). Let us note that a reconstruction of these criteria, which are historically mutable, should be also studied by a sociology of science along with a philosophy of science (although the epistemological criteria are basic for philosophy, while sociological mechanisms of the making of such criteria are sociology's main focus).

Actual value systems are not common to all scientists. They may agree with respect to the basic preferences which determine group ethos of scientists, but they will differ with respect to minor preferences. Value systems which differ with respect to some preferences will indicate different philosophical perspectives which, in turn, generate different classes of theoretical solutions. It may also happen that scientists linked to certain social classes will accept value systems with different principal values. Let us quote an example. Within the Marxist tradition there are basic differences with respect to preferences: in some state is the principal value, in some — society (sometimes a scientist identifies state with society and vice versa). Investigations conducted from the point of these, I think oppositional, values, have to reveal quite different statements and developmental predictions.

At the outset of social cognitive processes we are thus dealing with various groups of scientists who assume different value systems (which differ among each other with respect to principal or minor values) which, in turn, determine different philosophical perspectives. Each of these perspectives determines a certain class of solutions, and scientists choose the one which is based on their individual value system at the same time meeting the methodological conditions of the period.

The above simplifying assumption, namely that scientists produce independently of all pressures, guided by their cognitive aims alone is too crude and distorts the image of the functioning of science. Actually most of the scientists belong to some group or the so-called community. We are dealing with a mechanism of selection. It seems wrong to claim that in evaluating a given theoretical contribution a scientific group employs methodological criteria alone (ex. explanatory power). This is what Bell seems to think when he writes: "Sovereignity is not based on coercion, consciousness is individual and critical. As an idealized model it comes close to the Greek *polis* — republic of freemen, united in common striving for truth" (1976, 380). I do not think that a scientific group is guided only (or even mainly) by methodological criteria. They are accounted for, but they play a secondary role. The basic criterion of acceptance of any theoretical proposition is — to employ Kuhn's (1968) expression — its compatibility with the paradigm of the leaders, i.e. of those scientists who decide about means of producing and disseminating knowledge. Having those means they have actual chance to reject, disqualify, or keep silent about propositions which do not fit their paradigm. As we have already demonstrated (Buczkowski, Nowak, 1979, Buczkowski, 1979) the basic aim of the owners of means of producing knowledge is not to achieve cognitive tasks but to broaden the circle in which their paradigm is accepted. A scientist who represents a different point of view finds it hard to publish and disseminate his results, cannot win access to machinery, vital meetings, etc. These impediments cover a whole range of actions with administrative coercion as the ultimate weapon. As Kuhn (1968, 183) put it: "Acknowledgment of the only competent professional group in its role of a judge in professional matters has further consequences. Each member of this group must be — in view of their education and experience — considered the only experts on rules of the game or any other criteria of the acceptance of statements."

I think that it is thus legitimate to distinguish two basic groups of aims (preferences) in science. The first might be termed theoretical — i.e. striving for understanding and explanation of reality. The second group could be termed instrumental; it would include preferences of socially functioning scientific groups as a whole, striving to broaden the range of acceptance of group's leaders or of their paradigm. Theoretical aims characterize scientists as scientists, while instrumental ones — as human beings who have acquired "monopoly" for making science. In many cases instrumental aims can become the only aims pursued. This happens when a scientist has a theory which had been officially accepted or when he accepts and elaborates an official interpretation of the ruling concepts (or, in case of social sciences, of an ideological doctrine). His authority and social status supported by non-scientific means provide him with a chance to "shape" science according to his preferences. His domination includes recruitment of new scientists, decisions about research, etc. It often happens that scientists who have been granted this kind of "noble status" prevent the dissemination and acceptance of theories they deem false. Needless to say, even the most modern science is full of instances of this kind which impeded growth of theoretical understanding, sometimes for quite long periods of time. Needless to say, too, no scientist can have a monopoly of truth, even within the Marxist interpretation of the latter (Nowakowa, 1976). Each theory can be relatively true at best, and its accuracy is not determined by an acceptance or rejection in some groups but by its cognitive efficiency expressed in explanatory power, etc. There are many cases of professionals whose enclosure within their theory was motivated only by prestige claims, unwillingness to acknowledge their helplessness or a simple theoretical blindness. Many discoveries were made by individuals whose bonds to a given branch of science was relatively loose, young researchers who saw that older scientists with their conceptual schemes failed to notice or did not want to see. These are the problems of a rejection of the traditional order which we shall discuss while analyzing crises in science and in social order.

A number of theoretical solutions can occur within a single paradigm which is based on some value system. The latter determines theoretical perspective which allows for a class of possible solutions. At this point scientific correctness, explanatory power, prognoses, etc. take over. From the group of theories which strive for acceptance within a scientific group only those actually disseminate which meet methodological criteria of the time. Let us add that these historically mutable criteria determine the range of social functioning of science, i.e. rules of scientific conduct, verification methods, etc.

We have neglected one of the most significant mechanisms of cognitive practice — self-cenzorship. Scientists who recognize social criteria of acceptance strive to bypass the foreseen difficulties, especially if their results appear to have low chances of survival. They may, of course, give research up or hide their results. We shall not discuss these problems here, nor the problems of avoiding some topics in general. A scientist who recognized the criteria of acceptance and expects no support from his scientific group, can try to make his proposal acceptable without changing its core elements. The simplest, but easily deciphered way out is to declare his support for some paradigm. This form of proceeding enabled scientists to develop, although slowly, their branches of science during the Stalin era: they praised Marxism in their introductions, and left it there, producing "normal" science in the rest of their work. Traces of this procedure remain visible in social sciences, where a declaration of being Marxist seems to be viewed as a necessity no matter what theory one constructs. In the humanities — and to a certain extent in natural sciences (if this strategy is followed there at all) — the above procedure is not effective in the long run, since the theoretical proposals are decoded when disseminated, which may mean administrative sanctions against the author.

Declarations can be twofold: theoretical and institutional. In the latter case a scientist voices his support for an institution which regulates a given group (ex. academic science), while in the former case he declares his access to a theoretical community, by, let to say, suggesting a reinterpretation of the leading doctrine. The latter case is mostly linked to a school originated by some classic and to the interpretations which claim to be historically more accurate. The above situation has been conceptualized by D. Bell (1976, 383 — 384) who spoke of "double face" of scientific activity. "In a well-known sociological dichotomy *Gemeinschaft* is the opposite of *Gesellschaft*, the big impersonal society based on secondary bonds, regulated by bureaucratic rules and linked by a sanction of removal. Both *Gemeinschaft* and *Gesellschaft* taken together, describe modern science. There is a scientific community, recognition by peers who have valid achievements, a community with a charismatic power of producing and guarding selfless knowledge. There is also a professional society, economic enterprise of huge size, whose norm is to make society or enterprise useful." Economic motives are added by Bell to purely cognitive characteristics of a scientific activity. However, both these aspects of scientific activities are subjected to the aims of the whole community of scientists, who strive to broaden the range of acceptance of their paradigm. Kuhn stressed it (1968, 21) when writing "Normal science often supresses original discoveries if they undermine its basic achievements."

The above-mentioned historical accuracy of an interpretation allowing for the best available reconstruction of the model (master) idea is the best proposal for a scientist who wants his ideas accepted. It allows to break loose of the pragmatic limitations while avoiding a direct challenge to the ruling paradigm or a confession of theoretical helplessness. This happens when empirical data cannot be explained within the established paradigm. As Kuhn, again, puts it (ibid. 100): "The first attempts to solve the problem are undertaken according to the rules which follow from a paradigm. If one fails to achieve any results, numerous concretizations of the paradigm are introduced. Each is different, partially effective, but none is effective enough to be considered a new paradigm. As a result of the dissemination of these different versions (which are most often viewed as corrections *ad hoc*) the rules of institutional science become more and more complex all the time."

Another method of winning favours of those who decide about means of knowledge production is to declare adherence to a paradigm and to formulate a theory so that its difference is not easily noticed. Scientists may employ the same conceptual categories which the group accepts but modify their meaning and acquire better results which describe a given aspect of reality in a more precise manner. This happens if the decision-makers do not allow for any revisions. In other words, this is a strategy for a dogmatic scientific group. If new theories in old clothes better explain some aspect of reality and coincide with the need of a new class — the scientists who propose them become new owners of means of knowledge production and their new paradigm becomes an official one.

Apart from officially existing paradigms there may be scientists who play the role of "theoretical dissidents". Their theoretical works cannot be accepted by scientific groups for they definitely distinguish themselves from all official solutions and institutions. Their differences may thus be both theoretical and institutional (they disagree with an institutional manner of acceptance), or either theoretical or institutional. This group includes also those scientists who have been eliminated from the scientific community when scientific group failed to accept their work. Dissident concepts can be disseminated through mass media only when a sponsor can be found, i.e. either a group within a ruling class or a new class which strives to break the hegemony of paradigms defended by their political opponents. These are ideological matters, linked to the dissemination of consciousness contents in the whole society.

4. Functioning of science and ideological requirements

An acceptance of a theory by a scientific group does not mean this theory will be socially disseminated. Another sieve of selection is needed, i.e. from the point of ideological requirements of the ruling class. In societies which Marx analyzed the sphere of spiritual production was subjected to the material one. Subjection manifested itself in the requirement that theoretical concepts should secure the satisfaction of the interests of the economically ruling class. With a certain modification of Marxian theory one may say that this subjection will be controlled by every class which organizes social order, i.e. rules. In order to let a theory proliferate one has to recognize that theory motivates such social actions which stabilize the position of the ruling class in social structure. Things look differently in case of ideologically neutral sciences (ex. physics) which provide the rules of practical actions.

Motivational role of a scientific theory is expressed in dissemination of the values which are basic for the ethos of the ruling class and in their theoretical justification. It may easily be noticed that motivational role of a theoretical doctrine is linked primarily to its ideological dimension. Social dissemination is the fate of those theories which stabilize the existing social order best of all. This follows from the fact that in class societies the control over means of knowledge production depends on loyalty to the ruling class. Clearly all proposals linked to the dominated groups' ethos will be rejected, or, to be more precise, those, which assume values represented by these ethoses. Scientific group selection does eliminate also the proposals which harm the ruling class. My point of view is an interpretation of the classics from *German Ideology* (Marx, Engels, 1975, 50): "Ideas of the ruling class are in each epoch the ruling ideas, which means that the class which is the materially dominating force within a society is at the same time a ruling spiritual force. The class which controls means of material production controls also means of spiritual production, which makes it rule also over the ideas of those who lack means for spiritual production." Control of means of spiritual production was clearly evoked by them.

This quotation can have a twofold interpretation. Literal one claims that owners of means of material production are at the same time owners of means of spiritual production. If this interpretation is accurate, the classics assumed an idealized model of a society with two antagonistic classes. There is also an interpretation according to which the ruling class hands the control of means of spiritual production over to another group (in the Marxist tradition one calls the latter "literary representatives"), I call them controllers of means of spiritual production, requiring obedience and ideological legitimation of its interests. In both cases an individual scientist is subjected to the controllers of means of spiritual production. He is a scientist only in so far as his work and conduct correspond to the controllers' wishes.

Some Marxists claim that theory does not depend on ideology but results from a certain social practice. This seems to be the view of O. Cetwiński, who writes, for instance: "It is wrong to reduce ideological tasks to description and explanation. Ideology has also to justify some values and norms" (Cetwiński, 1979, 162). According to our concept of the group ethos, those explanatory and evaluating beliefs are the core of an ethos. An ethos in turn is produced by a group of scientists linked to some type of an activity. Social ideology is thus made of those elements of the ruling class ethos which are socially disseminated. It cannot be an autonomous creation of ideologies, who simply make use of theories or their elements, adding value systems to descriptive-explanatory stratum. Social efficiency of a theory which wants to be an official ideology in explaining and describing social phenomena is thus forced to meet the conditions of description and explanation. What matters is not an adequate description of reality. To the contrary, the better it obscures actual social mechanism, the better its chances to fulfil ideological functions. "Ideology strives for an essential falsity" (Nowak, 1984, 57), and the basic ideological theses have to turn human minds away from actual determinants of phenomena pointing out only some of them, which are of little significance. Ideology prevents the application of testing procedures, since it includes dogmatic statements, requires their acceptance and introduces ritual actions.

Considering global society from the point of class divisions two states can be distinguished. The type of a totalitarian society (Buczkowski, 1984) is most idealized and does not occur in pure form. In this society all forms of social activity have been subjected to the satisfaction of class interests of the ruling class. In such society an ethos of the ruling class would have to be identical to consciousness of all its members, both rulers and ruled. To achieve this state of affairs one would have a perfect dissemination technique and persuasion technology which would guarantee instant acceptance of all propaganda theses. *1984* by Orwell is a literary image of this type of society. Psychological efficiency of techniques he described cannot be analyzed within the frame of the present paper.

Historically existing societies approach the above model to a certain extent. Hitler's fascism or Stalin's system can be considered approximations of quasi-totalitarian societies. Quasi-totalitarian societies have inner mechanisms which allow other classes or groups to pursue their interests, too (I have no place to analyze these mechanisms — cf. Nowak,

1983). Let us limit ourselves to a general statement that a fully totalitarian society would result (after an elimination of all types of activities which are basic for society) in the annihilation of society (Buczkowski, 1984). Most of empirical societies allow for other class and group interests to be pursued to a certain extent, provided it does not clash with the pursuit of the interests of the ruling class or when the ruling class itself remains relatively weak. Thus mechanisms of disseminating knowledge and values in a society include most of group ethoses of the dominating groups. Empirically various pressure groups occur which modify contents of mass media (Gans, 1979, 249 — 278). The actual working of mass media respects, under the conditions of stability of social systems, interests of social classes which control the basic types of material means. This respect is manifested in dissemination of values (at least the basic one) and of their theoretical justifications which are characteristic for given ethoses. From the point of sociology of knowledge one should not only analyze techniques of communication but also discover relations between informational contents and elements of ethoses of the dominating groups. Let us also note that apart from the theoretical production of knowledge there are also other sources of potential knowledge which might be used by mass media or educational institutions — ex. tradition (literature, art), religion, et al. These are more detailed dimensions of a spiritual production characterized, I think, by similar mechanisms and ultimately dependent on science. Pursuing these problems would lead us too far from the topic of the present paper. Let us consider the relations between social actions and acceptance of disseminated values and their justifications.

5. Social consciousness and social action

Our discussion focused on an ethos of those groups which dominate in a society. This follows from an acceptance of a classical theoretical perspective in sociology of knowledge which concentrates on deformations or inadequacies of explanatory knowledge which is representative for the ruling groups. The problem of social consciousness of the dominated groups is then reducible to a study of dissemination of every respective ethos and its motivational efficiency as compared to actions undertaken on its basis. The concept of social class structure we had assumed so far rested upon an assumption that the basic criterion of class divisions is control of some material means and — as a consequence — of influencing beliefs which are disseminated in a society. Implicit assumption limited the disseminated ethoses to the ones belonging to the dominating groups alone. These elements of the ethoses which fulfil ideological functions compose the consciousness of dominated groups. It was thus assumed that dominated groups are unable to create particular ethoses. This is, indeed, the case, if we assume that science, which produces the bases for each ethos is in turn dominated by the ruling groups in the above-mentioned sense. However, this cannot be true with respect to ideologically neutral sciences, ex. mathematics, physics, etc. In their case the existing paradigms are subjected to inner scientific selection alone and they do not require another ideological verification to be socially disseminated. We note historical cases of class interventions into exact and natural sciences, but they were limited to quasi-totalitarian societies. In "normal" situations such interventions usually do not occur, unless a scientist starts an action outside of science. Hence paradigms of ideologically neutral sciences are socially disseminated in the same form they function within respective scientific groups. They are fragments of consciousness of the dominated groups. From the point of disseminating mechanisms all dominated groups accept these contents along with the ruling ones. Their dissemination is achieved through the educational institutions. Practically acceptance can be reduced to an acquisition of practical activities, indispensable in professional work, drawn from knowledge and values represented by these sciences. Not every production must be started with theoretical assumptions of new technologies in minds of direct producers, neither do they have to know nomological connections between states of affairs. Rules can be habitually employed without any profound theoretical reflection. Let us note that science and technology can fulfil ideological functions as well (cf. Habermas, 1970). It will do so, however, under the conditions described in a more general way above.

If elements of social consciousness of dominated groups include only practical rules, then common sense consciousness covers just these rules, as has been often pointed out in literature. Apart from rules acquired through education and socialization, there may be rules resulting from individual activity, either autonomous or shared in a peer group. Growing vegetables in a garden requires only practical rules (digging, sowing, watering), so does cooking, which also usually lacks theoretical justification in reflection on food processing and technology.

Acceptance of introductory remarks on social consciousness according to which its contents are accepted only by individuals who constitute social groups may suggest that my sole interest is with an individual action. An answer to a question about genesis of some actions (knowledge, values, attitudes) and predicted direction of behaviour are at stake. Let us point that the concept of social structure assumed in the present paper rules individualism out (and the latter is assumed in most sociological theories which discuss the problem of action). What sounds obvious in psychological theories, must give rise to doubts in sociological ones. In action theory an assumption of individualism manifests itself in a mechanical transmission of the principle of rational decision-making from individual to social scale (Nowak, 1984, 40) or a level of social groups. One assumes that consciousness of a decision-maker (knowledge and preferences) determines whether an action will be undertaken or not. Along with another assumption, which sees a group as a collective of similar individuals, they force us to characterize group action on the basis of a sum of identical or similar decisions. One often encounters the following scheme of explanation of effects of a group action: a group has common knowledge and preferences which allow for an undertaking of some action. If this action leads to the most prefered result, it is undertaken. However, nomological structure differs for a social group and an individual level. The above scheme holds true for an individual level, not for a group one. Explainig actions of integrated social group we should focus on objective interests — whether they are conscious or not decisively influences their pursuit. Ideology, which forms the basis of group interests has to disguise them. Therefore ideology may refer to various explanations which guide practical action of social groups. Let us consider an example. An individual capitalist enterpreneur can have various motivations in his activities: he may strive to produce something useful, create jobs for workers who lack capital, reinforce the position of his state, fulfil his deepest desires or increase his capital. Hypothetically each of these motivations can be ascribed to his actions. However, a global effect of the activities of the whole group of enterpreneurs is that the workers are being exploited no matter what their living standards are. This example illustrates difference of nomological structures in explanations of individual and collective actions.

In stable society social groups which are dominated and undertake some actions contribute to the pursuits of interests of dominating groups. The elements of the latter's ethoses are supposed to motivate individual to actions whose global effect will contribute to the satisfaction of class interests of the ruling classes, preventing individuals concerned from understanding the bonds between their actions and respective global effects.

Situation changes when the existing social order is being questioned as a result of class struggle. We shall not describe these situations (cf. Nowak, 1981, 1984) limiting our attention to the making of beliefs which question ideologies. L. Nowak uses the concept of utopia, which as opposed to an ideology, "aims towards an essential truth" (Nowak, 1984, 57). "If material interest of the oppressors requires a disguising of essential features of social reality, the interest of the oppressed requires that they be revealed (...). If the basic interest of the masses consists of acquiring control over means of production, the derivative interest of the masses includes recognition of everything which causes helplessness and prevents the pursuits of the basic values. The derivative interest of masses of direct producers thus consists of recognizing the essential aspects of social reality" (ibid. 56). The above mechanism holds for all groups which are subjected to domination. However, in revolutionary periods a society as a whole opposes the ruling class. It covers all dominated groups deprived of control over material means and also fractions of controllers who are in some way subjected to the ruling class. The latter enables society to use communication channels in order to disseminate utopian contents. It is essential for revolutionary situation to occur that mass direct communicatfon is established competing with mass media. Frequent contacts and common focus on struggle makes the beliefs, which are functional with respect to the actions masses undertake, proliferate very rapidly without previous communication channels. There are also very "simple" forms of dissemination — leaflets, wall graffiti, etc.

The first of the mechanisms which generate new beliefs and values consists of an employment of theoretical constructs created by scientists who broke out of the established paradigms (theoretical dissidents) — in this case we deal with ready-made system of beliefs which forms a basis of actions and motivates revolutionary social masses. This presupposes a situation in which a theory is based on values accepted by the masses. These values are most often opposite of values from the ethos of the ruling class, which had been adequately reconstructed. The second mechanism consists of generating a utopia which rationalizes actions which are being undertaken. Spontaneously appearing leaders verbalize the aims of the revolution and point to means of achieving these aims. Those two mechanisms correspond to two different courses of a post-revolutionary process. If ready-made construct is accepted it becomes a background for a new paradigm — and if victory has been won — of a new ideology. Current justifications — because of the numerous axiological elements — may give rise to later heresies which fight for the purity of revolutionary values. The former constructs, being systematic theoretical sets, are more susceptible to modifications through apparently conservative concretizations. Principal and justifying statements are completed with instrumental values and their theoretical justifications. The process of turning utopia into an ideology reveals gradual disguise of principal values and upward rise of instrumental ones with their respective system of knowledge. A developed ideology includes only instrumental values (plus their theoretical justification) and corresponding rules of social practice.

6. Concluding remarks

My aim in the present paper was to outline problems which, should one accept my definition of the subject matter of sociology of knowledge, form a certain interpretation of the Marxist thesis on "determination of social consciousness by material conditions of being". The scope of the present paper does not allow me to go far beyond the outline of the main directions of analysis. Mechanisms of social conditioning of the contents of social consciousness which had been discussed are, I think, the basic ones. In further analyses one should reject simplifying assumptions, both explicit and tacit ones, especially those we do not clearly perceive. However, even this bare outline allows, I hope, for operational formulae to be empirically verified. However, these steps would take me too far beyond the aim of this paper.

Translated by Slawomir Magala

BIBLIOGRAPHY

1. Bell, D. (1978), *The Coming of Post-Industrial Society. A Venture in Social Forecasting*, Penguin
2. Buczkowski, P. (1976), "The Marxian Category of a Bourgeois Scientist", in: *Poznań Studies in the Philosophy of the Sciences and the Humanities* vol. 2, no. 1
3. Buczkowski, P. (1982), "Toward a Theory of Economic Society", in: *Social Classes, Action and Historical Materialism, Poznań Studies in the Philosophy of the Sciences and the Humanities*, vol. 6
4. Buczkowski, P. (1979), "The Problem of the Division of Labour from the Point of an Adaptative Interpretation of Historical Materialism", in: *Ekonomista* no. 5 (in Polish)
5. Buczkowski, P. (1984), "Rationality and Levels of Social Organization", in: *Poznańskie Studia z Filozofii Nauki*, no. 8 (in Polish)
6. Buczkowski, P., Nowak, L. (1979), "Idealization and Significance. Case Study: Marxian Class Theory", in: A. Klawiter, L. Nowak (eds.), *Discovery, Abstraction, Truth, Empirical Data, History and Idealization*, PWN (in Polish)
7. Buczkowski, P., Nowak, L. (1980), "Werte und Gesellschaftsklassen", in: A. Honneth und U. Jaeggi (eds.), *Arbeit, Handlung, Normativität*, Suhrkamp, Frankfurt/Main
8. Buczkowski, P., Klawiter, A., Nowak, L. (1982), "Historical Materialism as a Theory of Social Whole", in: *Social Classes, Action and Historical Materialism, Poznań Studies in the Philosophy of the Sciences and the Humanities*, vol. 6
9. Cetwiński, O. (1979), *The Bases of the Theory of Politics*, PWN (in Polish)
10. Habermas, J. (1971), *Toward a Rational Society*, Beacon Press
12. Kmita, J. (1982), *On Symbolic Culture*, COMUK (in Polish)
13. Kolakowski, L. (1978), *Main Currents of Marxism*, vol. II, Oxford
14. Kolakowsk, L. (1979), *Main Currents of Marxism*, vol. III, Oxford.
15. Kolakowsk, L. (1982), "Epistemological Meaning of Sociology of Knowledge", in: *Can Devil Be Saved and 27 Other Sermons, Aneks* (in Polish)
16. Kuhn, (1968), *Structure of Scientific Revolutions* (Polish translation)

17. Mannheim, K. (1952), *Ideologie und Utopie*, Frankfurt/Main
18. Marx, K., Engels, F. (1975), *German Ideology*, in: K. Marx, F. Engels, *Works* vol. 3, KiW (Polish translation)
19. Merton, R.K. (1982), *Social Theory and Social Structure* (Polish translation)
20. Nowak, L. (1974), *Foundations of the Marxist Axiology*, PWN (in Polish)
21. Nowak, L. (1983), *Property and Power. Towards a non-Marxian Historical Materialism* Reidel, Dordrecht/Boston/Lancaster
22. Nowak, L. (1984), "Ideology and Utopia", in: *Poznańskie Studia z Filozofii Nauki* no. 8 (in Polish). (See also an abbreviated version in this volume)
23. Nowakowa, I. (1976, "Partial Truth — Relative Truth — Absolute Truth. An Attempt at the Reconstruction of an Ordering Concept of Essential Truth", in: *Poznań Studies in the Philosophy of the Sciences and the Humanities*, vol. 2, no. 4
24. Polanyi, M. (1951), *The Logic of Liberty* London
25. Schaff, A. (1970), *History and Truth*, PWN
26. Weber, M. (1946), "Science as a Vocation", in: Gerth, Mills (eds.), *From Max Weber*, New York
27. Ziółkowsk, M. (1982), "How Can One Make Sociology of Knowledge Sociological?", *Studia Socjologiczne* 1-2 (in Polish)

Leszek Nowak/Poznań

IDEOLOGY VERSUS UTOPIA
A contribution to the analysis of the role of social consciousness in the movement of socio-economic formation

The main purpose of the present paper is a concretization of a certain fragment of the theory of a historical process, introduced in earlier works. It is a concretization of the model of a socio-economical formation[1] according to the role played in this formation by the so called "social consciousness". In order to justify the inverted commas, however, we shall have to carry on with certain introductory considerations.

I. ON THE NATURE OF COLLECTIVE CONSCIOUSNESS

1. Materialism – institutionalism – idealism
It is beyond any doubt that in social life it is possible to distinguish three different areas: production (the economy), politics, and spiritual production (culture). Somehow less visible is the fact that these three spheres of social life are characterized by a similar internal structure.[2] In each of them it is possible to distinguish the material level (the basis), composed of a certain type of material means (the means of production, the means of compulsion, and the means of indoctrination) and appropriate social relations (those of production, power, and authority). Apart from the material level, in each of these spheres we can also distinguish the level of institutions (economical, political, or church — in the most broad sense, also nonreligious) and the level of collective consciousness (economical, political, and worldview).

Various theories of society stress the significance of different, as we shall speak of them later, social fields such as the material-economical one, the institutional-political one, the material-cultural one, etc., bringing the factors defined on these levels to the role of main factors of social phenomena.[3] It is possible to distinguish nine such fields, as the diagram below indicates.

Level \ Domain	Economy	Politics	Culture
Material	1.1	1.2	1.3
Institutional	2.1	2.2	2.3
Consciousness	3.1	3.2	3.3

And so, for example, Marxism considers the material level of the economy (1.1) as the leading social area; all other economical, as well as political and cultural, areas are to be explained with the use of factors defined upon the material-economical field such as effectiveness of the means of production and type of production relations. Liberalism, on the other hand, searches for the leading social field to be represented by the institutional level of politics (2.2); the fates of societies are dependent to a large extent on the type of political institutions, e.g. on whether they are democratic or authoritarian. A still different approach is e.g. the anti-naturalism of Spranger, which claims that the leading role in explaining the phenomena of the human world is performed by "the transcendental norms of the objective spirit" constituting a type of the logical structure of social consciousness. This may certainly be interpreted as a view which says that a leading social level is the level of worldview consciousness (3.3). The three approaches are expressions of, respectively, the material-istic, institutionalistic, and idealistic tendency in the theory of society. Each of these tendencies may at the same time occur either in a restricted or in a radical form. Marxism is an economist materialism and for that reason it is restricted. The author, on the other hand, supports the radical materialism, which assumes the entire material level of social life to be the leading social areas (1.1) — (1.3); in different historical epochs particular areas of this level attain a predominant position.[4] Similarly, in the restricted and radical versions there occur the views of institutionalism and idealism.

Let us add that the main opponents are materialistic and idealistic orientations. The point is that social institutions are nothing else but binding patterns of human social actions. The question concerning the basis of their existence may lead either in the materialistic direction — when it is pointed out that these models possess a normative power for people because in the last resort they are based on force, or in the idealistic direction — when it is acknowledged that at the basis of institutions lie the values believed in by people as well as their conviction that the so and so organized institutions lead to the realization of these values. At a closer inspection institutionalism requires either a material-istic or an idealistic support. Thus the problem may be brought to a question whether consciousness shapes, or whether it adjusts itself, to the material conditions of social life. Yet, on the basis of assumptions accepted in this paper the adjective "material" means more than "economical" — it equally refers to the material basis of the economy, politics, and spiritual production. It also assumes class divisions into those who have those material bases at their disposal and others like owners and direct producers, rulers and citizens, priests and faithful.[5]

2. Individual actions and collective actions

It seems reasonable to analyze at this point one of the theoretical sources of the idealistic view of society. It is represented by the transfer of the mechanism of rational decision-making, known to all of us from introspection, from the individual to the social scale. The point is that on the level of individual actions — if we disregard certain specific cases — it is really a fact that the factor which determines undertaking an action is represented by the decision-maker's consciousness: his preference and knowledge.[6] If this ascertainment is additionally supported by individualism, i.e. understanding human communities as groups of individuals, whose characteristic features are explainable by the features of these individuals, then it is easy to transfer the purposeful mechanism of action from the individual to the social level. In such a situation collective actions (e.g. the operation of the economy, or the operation of the state) are viewed as being determined by collective consciousness, quite similar to an individual action determined by the consciousness of individual subject. In this case ideal factors that comprise collective consciousness are claimed to attain the standing of the leading factors of social life.

If, however, individualism is rejected in favor of holism, i.e. a belief that at least certain actions of a community undergo laws which are different than laws governing the individual actions that constitute them, then the fact that (individual) consciousness is the leading determinant of actions on the individual level does not automatically cause that (collective) consciousness is the leading determinant of actions on the global level. It is materialism that maintains that the hierarchy of determinants of collective actions differs from the hierarchy of determinants of individual actions: consciousness, which is the main factor on the individual level is a secondary factor on the level of group actions. Here it yields to the material interest of acting communities, i.e. increasing the profit, and also to the multiplication of power or authority — to stop at the interest of ruling classes. And so in order to explain the decision of a particular capitalist or a politician, one needs to refer to whatever goal he has and what means can lead him to this goal. This kind of explanation consists in reconstructing the purposeful structure of individual actions[7] and, in fact, means the humanistic interpretation[8] of these actions. In order, however, to explain the global actions of the class of owners or the class of rulers one needs to refer to the interest of these social classes and so he has to determine what is it that a given community must do in order to increase to a maximum the global profit or the global sphere of ruling regulation in the given social conditions (which are determined mainly by the ability of the antagonistic class — i.e. the workers or the citizens — to resist). Thus this type of explanation concerns a social process, that is an entirety which, though combined of conscious and purposeful actions performed according to the principle of rational decision-making, itself is ruled by laws of quite a different type: objective like the laws of nature and in the same way necessary. For that reason the explanation of collective actions assumes the form of explanation used by natural sciences.

One could then say that the essential structure of the type of individual action is different from the essential structure of the type of collective action: in the former the main factor is the individual's consciousness, while in the latter — the interest of a respective community. The transfer from individual actions to collective activity composed of them is for that reason an essential transformation, since the change that occurs in the course of this transfer concerns the essence of the activity.[9] And if what we said is true, then the answer to the question of what is the source of the idealistic view of society mentioned in the introduction becomes clear. It consists in a false assumption that beginning with individual actions explained with the use of ideal factors (the consciousness of each particular individual) and combining these actions into collective activity, it is possible to preserve the same principle of explication and explain this collective activity with the use of the same ideal factors (the consciousness of the community). But this is exactly the way it is not. One may know nothing about what particular capitalists or politicians do want; it is sufficient for him to properly recognize the material interest of the class of capitalists or the machinery of power in order to obtain some approximate explanation. The recognition of interest is obtained through the analysis of social relations — i.e. mainly but not exclusively the relation of forces between the antagonistic classes — to which aim the analysis of the content of anybody's consciousness is not necessary whatsoever.

3. Class consciousness and the consciousness of a class

Collective actions are, for the above reasons, significantly (and thus also nomologically) different from individual actions of which they comprise: the main factor for the former is the collective interest and not the collective consciousness.

What is, however, the content of this "collective consciousness"? The type of community that is attributed with collective consciousness depends upon the accepted concept of social structure, and particularly upon what divisions of a society are accepted as fundamental. According to the point of view adopted here, the fundamental division of a society leads to distinguishing antagonistic social classes — the oppressors (owners, rulers, priests) and the oppressed (workers, citizens, believers), single or cumulated.[10] Derivative social categories from the so understood classes are class fractions (the categories instrumentally subordinated to them), strata, or the categories of their representatives.[11] In this case the main variety of collective consciousness is, according to the point of view accepted here, class consciousness, i.e. the consciousness of human communities which is formed as a result of an unequal relation between people to the means of production, the means of coercion, or the means of indoctrination.

Class consciousness consists of these convictions (cognitive or valuating) which motivate the members of this class to actions that comprise collective activities necessary to realize the interest of that class in the given objective conditions. Let us assume for example that the interest of the class of proprietors requires the change of production into a new type of management or else this class will gain a lower profit than that possible to gain with the reorganization of the structure of production. In such an objective situation the class consciousness of this social stratum includes the beliefs that motivate the proprietors to change the production; thus class consciousness will be of a reformatory character. If, on the other hand, it happens so that the citizens are lowered in the social scale and thus they present no serious danger to the class of proprietors, than the interest of this class requires only to consolidate the lowering. In such an objective situation the class consciousness of the authority is of a conservative character: it encompasses the beliefs that motivate particular rulers to maintain the existing relations between them and the citizens. And so on.

Generally, it happens so that in the given objective conditions (determined by the relations with the antagonistic class, the level of the material means staying at the disposal of a given class or to which it is subordinated, etc.) class consciousness consists in any case of the convictions which motivate a typical member of this class to actions which, together, comprise the collective activity necessary to realize the interest of this class in these conditions.[12] In other words, one could say that class consciousness comprises the convictions which are functional[13] in relation to the interest of this class in the given objective conditions.

The question is who believes in such a defined class consciousness. Usually no one. The above mentioned condition of the functionality of class consciousness determines the ideal type of consciousness (idealized consciousness), which does not have to be, and usually is not, fully believed in by any member of the given class. In other words, it usually is so that no member of the given class believes in these and only these propositions which meet this requirement of functionality in relation to the interest of his class in the given objective conditions. Thus he does not hold all of the beliefs functional in relation to his class interest and not all the beliefs he holds are — in the given objective conditions — functional to this interest. What one believes in is determined by a number of factors, the class interest, though a significant factor, is only one of them. Others are: the type of intellectual tradition with which an individual identifies himself, the influence of persons whom the individual considers to be authorities, the influence of propaganda, etc. All these circumstances determine that the consciousness of an individual belonging to a given class departs — to a different degree — from the one the individual ought to possess in order to realize in his or her actions the interest of the class to which he or she belongs, i.e. from class consciousness. If we define consciousness of a class as the set of beliefs held at a given time by the members of that class[14], then we could say that the class consciousness determined by the material interest of that class in the given objective conditions always departs — to a different degree — from the consciousness of that class in these conditions. For that reason class consciousness cannot be attributed directly to any member of the respective class. In certain conditions it is only possible to state that a sufficient approximation of class consciousness in a given objective situation is represented by the conceptions of the leaders or ideologists of that class.

An outstanding leader, or ideologist, is the one whose conceptions correspond to the objective interest of the entire class in particular conditions and whose way of thinking is followed — not without a certain difficulty and sometimes even tardily — by that of an average member of that class. The point is that although class consciousness always departs — to a different degree — from the consciousness of the respective class, there exists, however, a process of adjustment of the consciousness of a class to the objective interest of that class, or — in other words — the process of adjustment of the consciousness of a class to the class consciousness. This process occurs as a result of the activity of two mechanisms.

The first of these mechanisms is the elimination of individuals holding beliefs that are disfunctional to the class interest from the members of a given class. A capitalist who is deeply affected by the ideas of philantropy will give away his possessions and property to those who need it and thus will become a victim of competition. A member of the government who is deeply affected by the ideals of social self-government will give up his prerogatives in favor of the citizens over whom he is supposed to govern and, as a result, having lost his position, he will become a victim of his rivals. By and large, we can say that someone who holds beliefs that are disfunctional in relation to the interest of his class in the given objective conditions loses his position (his property, rule, or authority) within his class. And if he insists on holding these disfunctional beliefs for too long a time, he will simply be eliminated from that class.

The other mechanism is the learning of particular individuals — their giving up of disfunctional beliefs and accepting, instead, judgements which meet the requirement of functionality, i.e. the ones which motivate and promote actions that realize the interest of the class to which the individual belongs. Becoming convinced that holding certain beliefs is not worth the effort, people often give them up on any account — e.g. assuming them to be "notwithstanding with the reality". The one whose backbone is so strong that it makes him preserve his beliefs becomes eliminated on the principle of the first mechanism.

And so, the elimination of the faithful and the learning of the unfaithful promotes, as a result, the process of adjusting the consciousness of a class to the objective class interest. Since, in turn, class consciousness corresponds — by definition — precisely to the state of these interests, then this process could be otherwise defined as the adjustment of the consciousness of a class to its class consciousness. In this sense class consciousness is a border form of the consciousness of a class — the latter aims at class consciousness, though it never reaches it. This process lasts as long as the class interest remains in an unchanged form. When the social conditions (e.g. the relation between two antagonistic classes) change, and together with them also the form of social interest, a respective change affects also — by definition — the class consciousness. The empirical consciousness of the class then starts anew its process of adjustment — it is being adapted to the new state of the interests of its class or, in other words, it adjusts itself anew to the changed class consciousness. It has been illustrated graphically in Fig. 1.

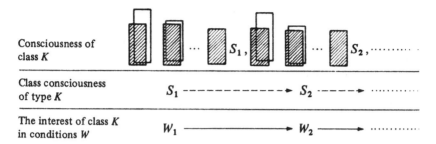

Fig. 1. Class consciousness *S1* (resp. *S2*, etc.) consists of beliefs which are functional in relation to the interest of class *K* in conditions *W1* (resp. *W2*, etc.) and only of such beliefs. The consciousness of class *K* in the subsequent stages of the process of adjustment overlaps more and more the class consciousness of *K*-type; in the border case they become identical. (Class consciousness is marked with the hatched area, while the part which is disfunctional in relation to the present state of interests of this class is marked with the white area). The process of adjustment, however, lasts as long as the class interest remains unchanged. The change in objective conditions, which leads to the new class interest of *K*-type, causes the process of adjustment to begin anew.

4. Consciousness as a secondary factor of historical development

If what we have just said is true, then it also becomes clear that the consciousness of a class (or any other human community) cannot perform a major role in a historical development. It is determined by the objective conditions, above all the relation between the antagonistic classes, rather than what the members of particular classes think of social or any other matters. It is the other way round: their thinking indicates a tendency of which we have already spoken — a tendency to adjust themselves to their material interests. In general — that is neglecting a small number of people who are always faithful to their beliefs — people think in such a way that helps them realize practically in the most effective way their interests, and this holds true for the both sides of the barricade of the class struggle.

For these reasons the transformation of the objective conditions of a certain type into the conditions of a different type (comp. Fig. 1) brings about the initiation of the process of adjustment changes of the consciousness of a class to the changed class interest. This transformation, in turn, occurs on the basis of principles that govern the class struggle — different in the case of the economic and political classes — which is the main factor among the factors causing the change in objective conditions. And if it is so, then in the first, fundamental model of the theory of a historical process one must abstract from the influence of consciousness upon social development. However, because this influence actually occurs in reality, it must be taken into account in some derivative model. This is exactly what I am trying to do in the present paper — to eliminate the initially introduced idealizing assumption which neglects the influence of consciousness[16], and to take this influence into account thus concretizing the model of historical development presented earlier. For quite obvious reasons the range of this concretization must remain limited to the theory of a socio-economical formation.

5. Social consciousness as a hypostasis

To be precise, the assumption eliminating the influence of consciousness upon the social development becomes weakened in this paper, i.e. it is being substituted by a logically weaker idealizing condition. In order to formulate this condition several explanatory remarks are necessary.

First of all, it is necessary to notice that the introduction of class consciousness to the model of a socio-economical formation is not, by any means, equal to or even followed by the necessity to take into account the "social consciousness". From the point of view presented here the "social consciousness" does not simply exist. The point is what should such consciousness be? Leaving primitive societies aside, only class societies (with single classes) or supraclass societies (with classes strengthened two or three times) can be found within "civilized societies". And as long as we are considering an isolated class society, there are no reasons to distinguish a "social consciousness" as a set of beliefs, which would be attributed to the society as a whole. The thing is that within the framework of an isolated class society there exists no material interest common for all the antagonistic class pairs: proprietors and direct producers, rulers and citizens, priests and the faithful. It is only the ideology of tyrant classes that deludes the oppressed that there exists the interest of "economy as such", mutual to the exploiters and the exploited, or the interest of "a state as such", bringing together the disposers of the mechanism of coercion and those who have no share in this mechanism, or — finally — the interest of "the development of culture as such", bringing together those who possess the means of indoctrination and those who have no possibility to materially strengthen their beliefs. Such common interests cannot exist in the most idealized model because the basic material interests of the mentioned pairs are antagonistic. Moreover — according to mechanism of adjustment of the consciousness of a class to the class interest, described above — these interests lead to the emergence of different, and in certain important parts quite contradictory, consciousness formations. Beyond these antagonistic consciousness of antagonistic classes there exists no "social consciousness" within an isolated class society. The same holds even more true for the isolated supraclass societies.[17]

And so the fact that we take into account the class consciousness of

both economically antagonistic classes does not necessarily have to mean
that we introduce a certain "social consciousness" because such a "social
consciousness" is a hypostasis on the level of the radically materialistic
approach accepted here. For that reason it is not so that first a certain
"social consciousness" is taken into account and then divided into certain
"forms".[18] The starting point in a two-class model of society is, on the
other hand, represented by two consciousness formations of both
antagonistic classes. Only in the course of the concretization of the class
structure of the society — the introduction of factions of classes, strata,
and representatives of the former — should it be necessary to introduce
appropriate types of collective consciousness generated by the interests of
these derivative social categories. The term "collective consciousness"
defines only the alternative forms of consciousness, from class consciousness, the consciousness of the fractions of classes, strata, and class
representations, to the consciousness of possibly further class-derivative
categories. The concept of collective consciousness does not, then, denote
any natural type,[19] but only a logical sum of such types distinguished on
the basis of the theory of classes and the above described mechanism of
generating consciousness by the material interest of an appropriate social
category. In the present paper, however, I shall limit myself to the basic
form of collective consciousness — the class consciousness.

6. Class consciousness in the socio-economical formation

Our assumption in the present paper is that the considered form of
collective consciousness is not the consciousness of a class but the class
consciousness. For that reason we shall assume, in other words, that the
process of adjustment of the consciousness of classes considered in our
model to their interest in every configuration of objective conditions has
reached the optimum state and the consciousness of a class is identical
with the class consciousness. Besides this weakening of the condition
neglecting collective consciousness, all the assumptions of the theory of
socio-economical formation — explained in the already quoted works —
are in force.

II. ASSUMPTIONS

1. Idealizing assumptions of the model of a socio-economical formation with class consciousness

The assumptions mentioned in the above title will now be presented and
discussed. It is accepted that the considered society is an isolated society
and for that reason all relations between that society and others are
neglected. The methodological sense of it is that no phenomena occuring

within the framework of that society must be explained with the use of variables referring to other societies.

Within the framework of the isolated society, an economic moment (of the field (1.1) — (3.1) of the table introduced at the beginning of this work) is additionally isolated, neglecting its relationships with the political moment and the moment of spiritual production (these moments are visualized by, respectively, colomn two and three of the same table). The so understood economical moment is a two-class society composed of economical classes only — owners and direct producers. For that reason, the two remaining class division into the disposers and non-disposers of particular material means, as well as state and church (in a broad sense) institutions being their superstructure and the political and meta-cultural consciousness, which adjusts itself to the interest of these classes, are neglected in this moment. The so understood economical moment is an idealization (ideal type) of a socio-economical formation, i.e. of societies (slave, feudal, and — to a certain degree — also capitalist), where the economical sphere really performs a dominating role. But even the economical moment itself will not be fully considered — we neglect here the role of economical institutions, i.e. field (1.2). It is only the material-economical level (field (1.1)), the theory of which is represented by model I presented in the already quoted works, and the consciousness-economical level (field (1.3)) — which is the subject of the present study — together with its influence upon the development of a socio-economical formation, that will be taken into account here. And so, as often as we shall speak here of class consciousness, we shall understand it as the economical consciousness of proprietors and direct producers, or — to be more precise — the class consciousness determined by the interests of these classes. The reason for it is that we accept here that the consciousness of these classes is identical to their class consciousness.[20]

2. Model I: the revolution of the exploited and the evolution of ownership relations

I shall now present in a very brief way[21] the main ideas of the model of a socio-economical formation, which takes into account only the material type factors, i.e. the model which is a starting point for the present considerations concerning the influence of class consciousness upon the development of the economical moment.

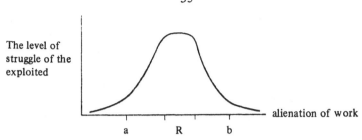

Fig. 2. The relationship between the economical struggle of classes and work alienation. Explanations: a - the threshold of class peace, R - the revolutionary interval, b - the threshold of declassation.

The starting point is the assumption that the level of class struggle of direct producers with the exploitation depends to a large degree upon the level of work alienation (the difference between the needs of the working class and the global variable capital invested by the class of owners in satisfying these needs). This relationship assumes the form of a bell curve — below a certain level (the threshold of class peace) of work alienation the class struggle stays on a low level. A similar situation occurs when work alienation crosses the threshold of declassation: the poor have no revolutionary inclinations. The revolutionary area, on the other hand, falls into the medium high values of alienation — when the exploitation is already painful but does not yet paralyze the social possibilities of resistance (Fig. 2.).

Let us now consider the course of class struggle in time. Let, in the starting point, the level of work alienation be below the threshold of class peace; the level of class struggle is at the same time relatively low. Each of the proprietors, however, aims at increasing his profits and, since the easiest way to do it is by taking them from the poor, the exploitation increases. Constantly more rigorous systems of appropriation (the division of the newly produced value into the surplus product and the variable capital) are spread because any of the owners who stays behind in the exploitation of his workers becomes wiped out by the competition. And yet, the point is that the result of it is an increase in work alienation, which in turn causes (comp. Fig. 2) an increase in the level of class struggle.

And so, work alienation increases up to the level of the revolutionary area. In this moment, at the time of revolutionary unrest, it appears that the class of proprietors is — in the conditions of model I — powerless. Private forces of coercion, which a particular proprietor is able to gather, are sufficient against local unrest at the time of class peace; they stop being sufficient, however, when the struggle of direct producers assumes a mass character. And one has to keep in mind that within model I there are no centralized forces of coercion, which the class of proprietors could possibly (on the basis of the alliance with the class of rulers) call on their side. Thus the only way out for this class is to make concessions — to change the relations of ownership in favor of direct producers.

As a result of the revolutionary stage, certain proprietors give away part of their ownership rights to their workers (e.g. they offer the workers ground lots on the condition of repaying or working the rent), thus increasing their production autonomy. Such proprietors are in this way turned into the proprietors in the new sense and the same refers to the workers. Thus besides the old classes of proprietors and direct producers — let us call them traditional — a new pair of antagonistic classes is generated — let us call them progressive, since they secure a higher autonomy for the workers. The new, progressive relations of ownership are spread as the time passes. The reason is that temporarily they ensure the proprietors for a social peace in their production units and perspectively they bring along higher economical effectiveness to them: a progressive labourer produces more effectively as he does possess a higher autonomy. As a result a constantly higher number of disposers of the means of production turns into the progressive class of owners and, at the same time, the old class division fades away. This results in a gradual disintegration of the old socio-economical formation which transforms itself into the new one based on different relations of ownership being not only more humane but also more effective.

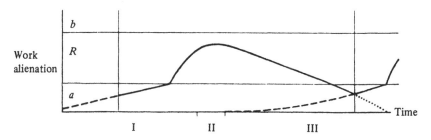

Fig. 3. The development of socio-economical formation (model I). The solid line marks the level of work alienation within the given formation. The intermittent line marks the level of work alienation within the progressive economic sub-system — it becomes a solid line when the progressive relations of a given formation become dominating relations in the following formation. The dotted line marks the level of work alienation in the traditional economic sub-system being residual within the new formation. I — the phase of increasing work alienation, II — the phase of revolution, III — the phase of evolution of ownership relations. The remaining symbols are explained in Fig. 2.

Thus we can say that there are three stages of the development of a socio-economical formation: the stage of the developing work alienation, the stage of revolutionary unrest, and the stage of the evolution of ownership relations. Such a model of the growth of an isolated economical moment without institutional and consciousness effects has been presented graphically in Fig. 3. This model makes in comparison to Marxian one the quite elementary historical facts more comprehensive. And so, economic revolutions break out in the central (rather than the final) stage of the socio-economical formation. They are not an indication of a contradiction between the forces of production and production relations, but rather of that of class antagonism. And first of all they are not a mechanism of an interformational transformation, which is of an evolutionary character and is carried out by the proprietors themselves, who gradually change their relations with direct producers. The historical role of economical revolutions consists instead in forcing the proprietors to accept this revision which improves the everyday life of the working class and also increases the effectiveness of the economy.

3. The duality of class consciousness: ideology and utopia

We shall now introduce to our model the factor of consciousness. It will be introduced to a limited degree: it will be restricted to only one form of collective consciousness, namely the consciousness of both antagonistic classes, and even this will have to comply with the assumption that this consciousness is ideally adjusted to the material interests of these classes in all the conditions that can be expressed in terms of the constructed model of a socio-economical formation.

Let us now consider what constitutes the consciousness of class (and at the same time, according to the assumption made, the class consciousness) of exploiters. Their basic interest is the maximization of the surplus product, while the secondary interest is expressed by every collective action taken by that class, which is a means for the realization of its basic interest. Since the fundamental factor counteracting the appropriation of the surplus product by the owners is the resistance of the exploited masses, then the (secondary) interest of the class of disposers of production means is that these masses understand as little as possible of the mechanism of the system to which they are subjected. For that reason they are not supposed to understand the very fact of the division of society, which runs along the line of controlling the means of production. Thus it is the interest of the class of proprietors to conceal the very criterion of class division. Moreover, the masses are not supposed to understand the antagonistic nature of this division, or — in other words — the fact that between them and the owners occurs a contradiction of interests consisting in the fact that the higher the profits of the proprietors are as a whole, the lower is the income of the class of direct producers as a whole. It is then the interest of the class of proprietors to conceal the antagonistic nature of class division. Finally, the masses are not supposed to understand that the only thing which can improve their social status is the open opposition towards the exploitation. They are not supposed to understand that it is only by way of struggle that they can obtain a certain production autonomy, that the prerogatives of ownership can be snatched away from the class of proprietors only by way of using force because in no other case shall the proprietors give them up. By and large, the interest of the class of proprietors consists in withholding from the masses the fact that their struggle is the fundamental source of historical progress. The masses are not supposed to be aware of all this because their getting to understand the material sources of class division, the contradiction of their interest with that of the class of proprietors, and the chances growing out of the opposition could embitter the class struggle, which even without it is the main obstacle for the realization of the principal interest of property. As far as the state of consciousness of the class of direct producers is concerned, the (secondary) interest of the class of proprietors is aimed, for the above reasons, at preoccupying the mentality of the masses with ideas leading them away from the recognition of the actual state of affairs. Such ideas can in other words be called significant-false ideas, which means[22] that even if they are true in the traditional sense, they still refer — at the most — to the secondary factors of the entire group of factors that determine the position of the masses. If the term ideology was to mean such a conception of social life which serves the interest of the class of oppressors (either economical, political, or spiritual) by suggesting to the antagonistic class a significantly false image of the given social reality, then one could easily say that it is the interest of the class of proprietors that the masses believe in the ideological image of the system to which they are subjected.

By way of digression we can also say here that the owners themselves believe in it: it is much more convenient for them to present themselves as honourable than as oppressors who live on someone else's work. Thus ideological thinking is imposed upon the direct producers, though it is usually imposed in a sincere way, almost always in good faith, i.e. in the way which corresponds to the authentic beliefs of the privileged. They themselves believe that the privileged position they have should be taken by them for granted because of their noble descent or as a compensation for their innovative and production organization talents, or simply because in the name of general happiness they realize the strategy of a "welfare state" and impose these beliefs upon the masses of country and city folk. And no wonder: it is much more convenient for them to perceive themselves in such a way than to see themselves as parasites living on the work of others or as exploiters profiting from the ill-paid work of workers, no matter whether the exploitation is done on the private or state ownership. And yet, these types of conceptualizations of the role of proprietors are significantly false and for that reason they are imposed upon the working masses. They constitute the ideological approaches to class systems, which at the same time express the social optics of the classes of proprietors and turn the attention of the working classes away from the content and essence of unjust social relations.

A contrast to the social thought of the privileged, i.e. ideology, is the social thought of the oppressed, which we shall call utopia. This term should not be understood in a depreciating way. On the contrary, while the material interest of the oppressors requires them to conceal certain important aspects of social reality, the interest of the oppressed requires these facts to be exposed. The important aspects of social reality are the ones which determine the detriment of the masses — incapacitation of workers in their work places, social poverty, or even complete poverty result from the monopoly of the private or state minority in the range of production means, as well as from the contradictory interests of the class of proprietors and the working masses. And so, while the basic interest of the masses is to gain control over the means of production, their secondary interest is the recognition of everything that causes their incapacitation and makes the realization of their fundamental aspirations impossible. Thus the secondary interest of the masses of direct producers consists in the recognition of significant aspects of social reality. It happens so in all "civilized societies" that the economical power of the minority is paid by the incapacitation of the majority in its work places. According to the economical theory of Marx this is the fact of a fundamental nature for the understanding of these societies. And so, it is the interest of the working masses to recognize the essential truth concerning the social life. In the past, these masses did try to reveal the essence of the system that exploited them by giving explanations which — though often presented in a mythological form — revealed much of this hidden essence. To give an example, let us quote the speach of one of the ideological leaders of the revolution of English yeomen in 1381:

"My fellow countrymen! Something is going on wrong in England and things will not be better as long as there is no unity of possession, as long as there are lords and their subjects, as long as all people are not equal. By what law do those who call themselves lords have authority over us? How did they earn it? Why do they keep us enslaved? If we are all the descendants of the same father and the same mother, Adam and Eve, how can they claim and argue that they are entitled to more rights than we are? Maybe because we are the ones who work and produce what they eat? They are dressed in velvets, purple and fur coats, and we wear but rugged linen. They have wine, roots and white bread and we get only rye bread, bran, straw, and water. They live in palaces and castles, while our fate is the pains and misery in the cold and windy weather. And yet, they owe their glamour only to our work. Stil, they consider us their servants and punish us when we do not obey their orders."[23]

As we can see from the above text, not only the fact of class division is recognized here but also the fact of the appropriation of the surplus product by the class of owners, as well as the contradiction of class interests. And so there are many elements of a class society that are revealed by that mutinous speach. A close look on the social system from the low level of society is much more cognitively fruitful than the one from the level of the privileged class. Social utopia always aims at essential truth, while ideology runs in the direction of essential falseness. It is probably for that reason that the social sciences are still able to explain only so little, because the scientists usually take the point of view of the privileged. And no wonder: it is the high levels of the society that offers money for the cultivation of science.

And so utopia aims at revealing the essence of the systems which harm a large majority of the society; it aims at revealing the essential truth. Another characteristic feature of utopia is that relates the materialistic explication of the present time with the idealistic approach to the future. The elimination of the sources of the present social evil is to automatically lead to the generation of a new system which appears as a negation of the opposed reality. Utopia offers no guarantee of a constitution of total freedom, equality, and fraternity but one: the good will of the revolutionaries. The language of interests, contradictions, and antagonisms in the approach of an existing society yields to the language of ideals and the assumption of the good will of the winners in realizing these ideals whenever the subject of the utopian thinking is the future. Utopia, then, relates the materialistic explication of the present time, aimed at obtaining the essential truth, with the idealistic program. And so, it realizes in a better or worse way the interest of the oppressed, but it does so only as long as the social system to which it is opposed lasts.

Thus the component of the consciousness of the class of proprietors is ideology, while the component of the consciousness of the working class is utopia. These elements differentiate — to a different degree, as we shall see, in the course of the socio-economical formation — the consciousness structures of both antagonistic classes.[24]

III. IDEOLOGY AND UTOPIA IN THE MOVEMENT OF A SOCIO-ECONOMICAL FORMATION (MODEL II)

1. The influence of ideology and utopia, taken separately, upon the struggle between economical classes

Let us now try to specify the corrections that need to be introduced to model I in relation to the admission of the influence of the consciousness of both economical classes. It is self-evident that both ideology and utopia influence the level of class struggle. The direction of that influence is also self-evident. Ideology, presenting a false picture of a system and blurring its class structure and the antagonism of interests, counteracts the class struggle, while utopia, stressing in a less or more evident way the elements of the world in which the masses live that are harmful to these masses, thus pointing out the enemy and also presenting the image of an ideal future, additionally motivates to fight the existing order. There also occur — to a certain extent reverse — a connections between work alienation and the susceptibility of the working class to ideology and utopia, but for the time being let us concentrate on the first relation.

We shall begin with an analysis of the influence of ideology and utopia (separately) upon the level of class struggle. In order to make it more clear it must be added that the mentioned relationships will be defined here exclusively according to the type. Therefore, talking about the influence of ideology upon class struggle, we shall refer only to the definitional features of ideology and not to the content of any ideological structure. For each particular ideology there exists a certain relationship of the mentioned type, defining its influence upon the development of a particular formation. Needless to say the establishing of the form of such a relationship is very difficult. It is possible that the entire problem is similar to that of quantitative relationships, in which the specification of a general mathematical formula is sufficient to establish the type of relationship, while in order to establish precisely the course of the curve an empirical evaluation of numerical constants is necessary. In our case, an equivalent of the evaluation of these constants would be the analysis of the relation between the content of a particular ideology and the mass actions. Only such an analysis would allow for a selection of a particular relationship out of the set (type) of relationships defined on the basis of general considerations. Similar remarks concern the influence of utopian thinking upon class struggle and also — by and large — all the relationships in which the degree of spreading of consciousness formation is a determinant.

The spread of ideology (or, resp., utopia) influences the level of class struggle in such a way that it lowers the level of class struggle in the case of ideology and raises it in the case of utopia. For that reason the mentality of direct producers enslaved by ideology decelerates the readiness of the masses to be resistent on a large scale. This may be interpreted as a transposition of the threshold of class peace towards the area of revolutionary unrest — the value of the threshold of class peace is thus increased. It also shortens the area of revolutionary unrest and accelerates the atomization of the masses. This last effect may be interpreted as a transposition of the threshold of declassation in the direction of the revolutionary area, which means lowering the value of the threshold.

The ideological motivation of direct producers, i.e. the influence of social utopia, exerts quite a different influence upon the level of class struggle; for the time being we also assume that it is constant. Revealing and defining the enemy, as well as contradicting the hateful reality with a vision of an ideal system, social utopia strengthens the tendency of the masses to fight. It can be expressed in such a way that it lowers the value of the threshold of class peace, lengthens the area of revolutionary unrest, and increases the value of the threshold of declassation. The level of class struggle corrected by the (constant) influence of the spread of ideology (or, resp., utopia) can be presented in the way shown in Fig. 4.

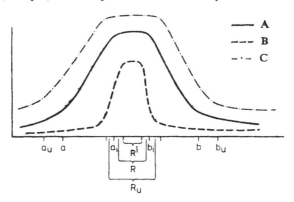

Fig. 4. The relationship between class struggle and work alienation in the conditions of the lack of any ideological doctrine (*A*), with the (constant) influence of ideology (*B*) and (constant) influence of utopia (*C*) taken into account. Let us add that this diagram specifies the type of correction of curve *A* according to the (constant) influence of ideology (utopia). The position of curve *B* may be higher or lower depending on the influence of particular ideological contents upon the masses; however, it must always run below curve *A*. Respective remarks refer to curve *C*, which runs above *A*. The influence of ideology and utopia (assuming the constant character of popularization) upon the level of class struggle can be presented as follows: points a_u, a_i mark, respectively, the thresholds of class peace: the threshold lowered from (the value specified for) a to a_u and the threshold raised from a to a_i under the influence of the popularization of, respectively, utopia and ideology. Indeces u and i in the remaining usages have a similar sense.

2. The susceptibility of working masses to ideology (or, resp., utopia).
The commonness of the belief in a particular ideology (or, resp., utopia) among the masses is not, in fact, constant but depends on work alienation.

The most effective influence that ideology exerts is that upon the direct producers who are well to do — if the system satisfies their needs to a high degree, then the working classes, also to a high degree, accept the ideological thinking behind which the class of proprietors conceals its material interest. When work alienation reaches the revolutionary area, the popularization of ideology reaches the lowest level. It raises together with the atomization caused by indigence but it does not reach the initial level of ideological indoctrination. And so, below the threshold of class peace the popularization of ideology among the masses is at its utmost, in the area of revolutionary unrest it assumes the lowest values, and it reaches the medium values above the threshold of declassation (Fig. 5). In the case of utopia it is the other way round — below the threshold of class peace the popularization of utopia is the lowest, it is the highest in the revolutionary area, and it assumes the medium values above the threshold of declassation (Fig. 6).

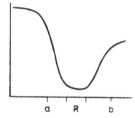

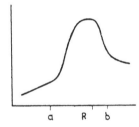

Fig. 5. The susceptibility of the working class to ideology depending on work alienation.

Fig. 6. The susceptibility of the working class to utopia depending on work alienation

3. The mutual influence of ideology and utopia upon the level of class struggle
Let us put together the influences discussed above: both considered ideological systems, ideology and utopia, influence the level of class struggle and the level of their popularization among the working masses depends upon the level of work alienation. Taking these influences into consideration, we can now establish the mutual influence of ideology and utopia upon the level of class struggle. The lower the popularization of utopia, the higher the popularization of ideology and *vice versa.* For that reason, for the low values of work alienation, where the number of

advocates for ideology is a lot larger than the number of advocates for utopia, almost such effects take place which are characteristic of the influence of ideological indoctrination of masses upon the course of class struggle — raising the value of the threshold of class peace and lowering the level of class struggle itself (Fig. 4). Above the threshold of class peace, together with the increase in work alienation certain other effects start to appear. These effects are characteristic of the influence of utopian thinking of the masses upon the course of class struggle and include the expansion of the revolutionary area and raising the level of class struggle itself (Fig. 4). In this area the number of workers who yield to the influence of utopia is so large in comparison with those who are constantly influenced by ideology that the latter effects exert almost no weakening power upon the effects of the influence of social utopia. On the other hand, for still further values of work alienation the effects of ideology and utopia start neutralizing each other, since declassation brings ideological discord among the masses. As a result almost nothing changes in standard picture of the course of the curve of class struggle (comp. Fig. 2). The mutual influence of ideology and utopia upon the course of the curve of class struggle can, therefore, be presented in the way shown in Fig. 7.

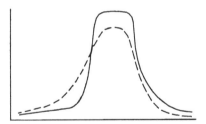

Fig. 7. The mutual influence of ideology and utopia upon class struggle. The intermittent curve marks the course of class struggle without the influence of consciousness

4. The influence of ideology and utopia upon the development of a socio-economical formation

Having established the way in which ideology and utopia influence class struggle, the correction of the movement of a socio-economical formation from model I presents no special difficulty any more. All three stages (of developing alienation, revolutionary unrest, and the evolution of ownership relations) of model I undergo certain modifications in the presently analyzed model II.

The stage of developing alienation is extended because the threshold of class peace is raised. This is often explained by the fact that at the first stage of the formation, when the level of work alienation is relatively low, the susceptibility to the ideological indoctrination is — for that reason — high (comp. Fig. 4). There will always happen to be a certain ideological doctrine, which — in a better or worse way — will perform the role of the autorationalization of potentates and blurring the nature of the system to the exploited. This does not have to be a doctrine in the sense of a systematized set of beliefs presented by someone who is professionally preoccupied with it; it can simply be the "thoughts of the ruling class", to refer to a well known statement by Marx, that "become the ruling thoughts".

However, it does not happen so throughout the course of the entire formation that the thoughts of the class of proprietors are the ones ruling over the mentality of the working class. As a matter of fact it happens so only in the stage of developing alienation. In the central stage of the formation it is the other way round — the thoughts of the working class are its own thoughts. Popularized at this time is social utopia, which strengthens and extends the period of revolutionary unrest. This, in turn, accelerates concessions in favor of the class of direct producers and thus it strengthens the process of forming of a progressive economical arrangement. That means it actually accelerates the evolution of owner- ship relations. Different utopias increase to a different degree the readiness of the masses to fight the exploitation because they give better or worse explanations of the harm to which the masses are exposed, and are at the same time less or more persuasive. Thus different utopias constitute better or worse bases for the programs of class struggle. To a varied degree every utopia activates the masses, and so every utopia intensifies and extends the revolutionary unrest, which further results in the acceleration of the evolution of ownership relations.

This very fact, however, is the reason for which the original ideology of the class of proprietors loses irrevocably. The split within the so far uniform class of proprietors into two competing in a less or more intense way factions, the interests of which are — to a certain degree — different, causes the decrease in the demand for the return of the original ideology, which used to rule over the masses back in the stage of developing alienation. Firstly, in the minds of the masses this ideology is to a large degree superseded by utopia. Secondly, its restitution is demanded by only part of the proprietors — those, who continue the traditional ownership relations, i.e. the ones which were once so effectively defended and excused by this ideology. Moreover, this part of the proprietors is less expansive and constantly shrinking. Thus the traditional ideology is somewhat revived in the stage of the evolution of ownership relations after the ideological disaster that occurred to it in the mentality of the working masses during the revolution in the center of the formation. However, soon it reaches its new level and, from this time on, loses its influence irrevocably.

Having changed the ownership relations, the progressive class of proprietors is bound to find itself a new ideology to justify its actions. It cannot use for that purpose any appropriate interpretation of the traditional ideology because it is used by its rival — the traditional class of proprietors. It must, then, be a completely new ideology. At the same time, the ideological situation in which the progressive class happens to be in the beginning of the stage of the evolution of ownership relations differs significantly from the ideological situation of the class of proprietors in the beginning of the formation. In that period, according to our assumption, dominated the ideological vacuum with a high susceptibility of a relatively well to do working class to the structures of ideological thinking. Now, at the beginning of the stage of the evolution of ownership relations, the mentality of the people stays under an overwhelming influence of utopian thinking, which is a heritage of the stage of revolutionary unrest. Thus the most convenient solution for the class of progressive proprietors would be the revindication of social utopia by way of such a reinterpretation of it which would deprive it of whatever in it speaks against not only the traditional but also the progressive ownership relations, or even against any type of ownership of production means limited to the minority of the society, on the one hand, and — on the other hand — an interpretation which would somehow justify such progressive ownership relations. In this way the adaptation of the reinterpreted utopia to the demands of a new, constantly growing in importance class of progressive proprietors is carried out. Utopia becomes ideology. And when the new, progressive ownership relations force out the traditional ones, when the old formation turns into the new one, the utopia of the masses from the previous formation becomes the ruling ideology of the new formation. At its origins, the majority of working masses is again seized with the hallucination of ideology (Fig. 4). The only difference is that the new ideology is a transformed utopia of their class predecessors.

Evidently, these processes are of an adaptational type, which means that they are statistical in their nature. There remains a certain group of supporters of the old utopia in an unchanged form preserving, in particular, the opposition to all rights of any minority to the monopoly of production means. And so the old utopia, the heresy, still exists equally with the utopia reinterpreted to the role of ideology. The supporters of the former call for studying canonical books constituting common sources of the ideologized utopia of new masters and their own doctrine in order to show that their orientation, now being forced out aside, is the true continuation of these books, while the officially proclaimed doctrine is a renegation.

Therefore, in the final stage of the socio-economical formation a dissonance begins between the ideology of the new class of proprietors and the heresy. Although they both grow out from the same ideological system, they differ radically in respect of the social sense. Heresy constitutes a ready embryo of a new utopia, the demand for which is indicated by the masses at the revolutionary stage of the new formation. To be more precise, heresy is one of the alternatives of utopian thinking, which stays at the disposal of the masses. Whether it wins in the competition or lose to the new alternative depends on which of them responds better to the new demand of the working class in the changed ownership relations.

On the other hand, the result of the ideologization of utopia which occurs in the stage of the evolution of ownership relations and from which objective processes benefit is the further acceleration of the process of changing into the new ownership relations. The point is that by verbal standards of people's utopia the progressive class of proprietors gains a new, effective device of ideological stupefying its working masses. For that reason their ability to resist is weakened and the progressive class wins social peace among its production units. On the other hand, by using traditional ideology — discredited at the time of revolutionary unrest — the class of traditional proprietors further antagonizes its direct producers with this factor. The distance between the progressive and traditional property on the scale of social peace is, then, further deepened, which constitutes another significant factor of a faster turning of traditional proprietors into the new ownership relations. The above described consciousness processes, the split of utopia into the new ideology and heresy, further accelerate the passing from the old formation to the new one. These processes may be graphically described in the way shown in Fig. 8.

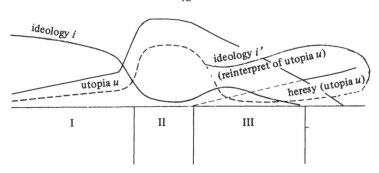

Fig. 8. The development of a socio-economical formation in model II. The thin curve indicates the level of class struggle and thus already includes the ammendments resulting from the activity of class consciousness (i.e. ideology and utopia). In order to make it simple the curves illustrating the spread of particular ideological forms in the course of the formation have also been included in the diagram. The thick solid curve illustrates the spread of ideology — the old one and the new one — into which utopia (the spread of it is marked with a thick intermittent curve) is transformed in the final stage of the formation. Comparing the picture of the curve of class struggle subjected to a doctrinal correction (the above figure) with the picture of the curve of class struggle independent of the influence of consciousness factors (comp. Fig. 4), we can see how insignificant are the corrections caused by these factors: the threshold of class peace is displaced (thus the stage of developing work alienation is extended), the level of class struggle in the revolutionary stage is raised — while the revolutionary stage itself is extended, and the stage of the evolution of ownership relations is shortened.

This basic outline: the transformation of utopia into the new ideology and, at the same time the emergence — as an opposition to it — of heresy continuing the original form of utopia, though too much idealized,[25] can be found with a certain approximation in many consciousness processes. A typical example is the development of the christian social doctrine in the ancient times and the middle ages. Initially, it was the utopia of the disinherited and exploited, including not only a very strong criticism of private property but lso a program of establishing communities of the faithful based, e.g. upon common ownership. The first fathers of the church were still preaching the following:

"You should grant everything to your brothers and not speak of property, because if we are all brothers as far as spiritual goods are concerned, the more we are brothers as far as temporal material goods are concerned" (Barnaba);
"Nothing stands out against the force of opulence, everything bends to its lawlessness... Are you not greedy or a thfef? Do you consider your property what you received under your management? The bread you are hiding should be given to the hungry, the clothes hidden in your chest belong to the naked, the shoes rotting in your house should be given to those with bare feet, and the treasure buried in the ground by your greed belongs to the one who needs it. (...) Should the people, who have minds,

be more savage than wild animals?! The animals use the fruit of the earth equally: the sheep graze on the same mountain pasture, horses eat grass from the same meadow, while we consider our exclusive property everything that ought to belong to everyone" (Basil the Great);
"We christians use everything that comes from God together and no one is excluded from using His gifts and benefits, as the entire human kind should equally use the good of God's mercy" (Cyprian), [26]

Later, the christian social doctrine was gradually affected by ideologization until Saint Thomas Aquinas, whose interpretation allowed for a private property under certain conditions (if the goods produced by it are treated as a help in doing good in order to be saved). Moreover, the tendency to gain temporal goods should, according to Saint Thomas, be adjusted to the appropriate status to which the individual belongs and to the functions which he or she performs within the society in relation with that status. Only the tendency to obtain more goods than it is suitable according to the status is reprehensible and indicates greed. On the other hand, the division itself, including the division into a peasant and noble, is a result of unequal human abilities and talents and, for that reason, different advantages offered by various people, from which the given social organism can benefit.

"For that reason Thomas Aquinas did not consider the organization of a society based on social and professional divisions as a result of a sin but rather as something desired by God in accordance with his idea of mercy. Nobody must try to cross the borders of his status or deny the vocation which he inherited from his father".[27]

Christian organicism, which sanctions the feudal class structure, has been expressed by one of the church's political writers in the following way:

"The health of the community will be secured if the higher members respect the lower members, while the lower members show the same relation with the higher ones".[28]

This indicates an almost complete disappearance of the spirit of the criticism of private property present in primitive christianity. By reinterpreting its sources, the christian social doctrine has taken a long way from utopia to ideology. This has borne fruit in the many heresies which — with all the differences between them — commonly referred to the original ideals of apostolic poverty, condemnation of riches, and the apology of common property. These heresies "all equally negate those principal values on which the backbone of early feudal societies is based and all oppose the individuality of man, liberated from the ties that bind him, to the authority, established traditionally and united with the system".[29] These heresies were the main source of inspiration for the anti-feudal social movements.

Translated by Krzysztof Sawala

NOTES

1. This model has been presented in my article, "Adaptation and Revolution", *Poznań Studies in the Philosophy of the Sciences and the Humanities*, vol. 6, 1982.
2. More details in L. Nowak, "The Notion of Material Momentum", *The Polish Sociological Bulletin*, 1980, no. 4 and P. Buczkowski, L. Nowak, "Werte und Gesellschaftsklassen", (in:) A. Honneth und U. Jaeggi, *Arbeit, Handlung, Normativität*, Suhrkamp, Frankfurt/Main 1980.
3. The explication of the phrase "the main factor for the class of magnitudes" is included in the article by L. Nowak, "Historyzm metodologiczny w kategorialnej interpretacji dialektyki" (Methodological Historicism in the Categorial Interpretation of Dialectics), (in:) Z. Cackowski, J. Kmita (eds), *Społeczna funkcja poznania* (The Social Function of Cognition), Warsaw/Gdańsk/Kraków/Wrocław 1979.
4. More details in my article, "Historical Momentum and Historical Epochs", *Analyse und Kritik*, 1979, vol. 1.
5. ibidem and also works quoted under 2.
6. More details in J. Kmita, L. Nowak, *Studia nad teoretycznymi podstawami humanistyki* (Studies in the Theoretical Foundations of the Humanities), Poznań 1968, chap. II.
7. J. Kmita, *Z metodologicznych problemów interpretacji humanistycznej* (Some Methodological Problems of Humanistic Interpretation), Warsaw 1971, chap. I.
8. ibidem.
9. This type of intuitions is certainly meant by S. Rainko, who claims in his book *Świadomość a historia* (Consciousness and History), Warsaw 1978, that on the level of actions consciousness is a determinant, while on the level of macrostructures it is determined by being. If, however, one agrees with the formulation of the thesis of pancausalism, saying that for each factor there exists its essential structure (the hierarchy of determinants), then Rainko's claim does not evidently express its meaning — everything has its determinants and determines something, among others, individual consciousness and socio-economical conditions. What I think the point is has been expressed in the formulation included in the text, using the conceptual mechanism explained in my article "Struktura, kategoria przedmiotowa, postęp" (Structure, Object Category, Progress), (in:) *Konfrontacje i parafrazy* (Confrontations and Paraphrases), *Poznańskie Studia z Filozofii Nauki*, 1979, vol. 4. Also compare my *Wykłady z filozofii marksistowskiej* (Lectures on Marxist Philosophy), vol. II: *Ontologia i epistemologia* (Ontology and Epistemology), Poznań 1977, chap. II.
10. Comp. works quoted under 1 and 4, while more details can be found in my book *Property and Power. Towards a non-Marxian Historical Materialism*, Reidel, Dordrecht/Boston/Lancaster 1983.
11. For information on categories secondary to economical classes comp. A. Jasińska, L. Nowak, "Grundlagen der Marxschen Klassentheorie. Eine Rekonstruktion", (in:) J. Ritsert (hrsg.), *Zur Wissenschaftslogik einer kritischen Soziologie*, Frankfurt/Main 1976, Suhrkamp p. 175 and next.
12. I refer here to the conception outlined in my book *U podstaw marksistowskiej aksjologii* (Foundations of Marxist Axiology), Warsaw 1974, p. 8. For more details in

the spirit presented here comp. P. Buczkowski, L. Nowak, "Werte...", op. cit.

13. I have taken the term "functional" from J. Hochfeld, *Studia o marksowskiej teorii spoleczeństwa* (Studies in the Marxist Theory of Society), Warsaw 1963. The concept of functionality, which I am using here, refers, however, not only to ideology — as in the case of the above mentioned author — but also constitutes a definitional property of all collective consciousness.

14. I refer here to a well known distinction introduced by G. Lukács, *History and Class Consciousness*, Cambridge Mass. 1968.

15. Comp. the works quoted under 10.

16. Comp. the work quoted under 1.

17. Only after the assumption of isolation is eliminated, a certain unity of interests of classes antagonistic to the external danger appears; in this way a common national consciousness is formed. This, however, concerns the intersocial relations, while the concept of social consciousness clearly implies the thesis of the existence of the unity of attitudes or views within the society, irrespective of the relations with other societies. Cf. my work "Zagadnienie narodowe w nie-Marksowskim materializmie historycznym" (The National Problem in non-Marxian Historical Materialism), Poznań 1981.

18. This, among other things, is what differs the conception presented in the text from that of J. Kmita, *Szkice z epistemologii historycznej* (Essays in Historical Epistemology), Warsaw 1979, and by the same author "A Humanistic Coefficient of Activity and the Psychology — Humanities Relation", (in:) J. Brzezinski (ed), *Consciousness: Methodological and Psychological Approaches, Poznań Studies in the Philosophy of the Sciences and the Humanities*, vol. 8, 1985. These two conceptions are similar in respect of the ontological status attributed to non-individual consciousness, a similarity which has already been mentioned (comp. reference 12).

19. The concept of a "natural kind" can be interpreted as an *F*-kind in the understanding of the categorial interpretation of dialectics (comp. the works quoted under 9). A wider explication is included in the article by G. Banaszak, "Zagadnienie klasyfikacji w kategorialnej interpretacji dialektyki" (The Problem of Classification in the Categorial Interpretation of Dialectics), (in:) *Konfrontacje i parafrazy* (Confrontations and Paraphrases), *Poznańskie Studia z Filozofii Nauki* 1979, vol. 4, p. 39 and next.

20. Many additional effects of economical, demographical, etc. nature have been neglected here. On this matter comp. the works quoted under 1 and 10.

21. ibid.

22. I. Nowakowa, *Z problematyki teorii prawdy w filozofii marksistowskiej* (On the Problems of the Theory of Truth in Marxist Philosophy), Poznań 1977, chap. III.

23. Quoted after M. Beer, *Historia powszechna socializmu i walk spolecznych* (Polish translation), (The General History of Socialism and Social Struggles), Warsaw 1957, vol. I, p. 301.

24. Needless to say, the consciousness of a class (and even class consciousness) cannot be reduced to utopia or ideology. It contains appropriate practical knowledge (e.g. technical-economical knowledge in the case of economical classes) as well as a certain structure of a higher rank, worldview, which combines these three components (ideology, knowledge, utopia) into a uniform consciousness formation.

25. Many simplifying assumptions have been taken here for granted. For example, it is taken for granted within model II that there exists only one ideology and one utopia, which are generated at the starting point of the socio-economical formation and which last throughout it. Obviously, it does not have to be so and it usually is not. If this condition was to be rejected, then many ideological or utopian doctrines which emerge and possibly disappear at different moments of the development of the

formation would have to be introduced, while the thick lines in Fig. 8 would consist of fragments of these particular lines representing particular doctrines in the state of maturity, i.e. when they managed to gain the strongest support.

26. Quoted after M. Beer, *Historia...*, p. 153 — 155.

27. H. Becker, H.E. Barnes, *Rozwój myśli spolecznej od wiedzy ludowej do sociologii* (Polish translation), (Social Thought from Lore to Science), Warsaw 1964, p. I, p. 340.

28. This a comment of John of Salisbury quoted after R.H. Tawney, *Religia a powstanie kapitalizmu* (Polish translation), (Religion and the Origin of Capitalism), Warsaw 1963, p. 50.

29. S. Czarnowski, *Odnowienie kultury* (The Renewal of Culture), (in:) *Dziela* (Works), vol. 1: *Studia z teorii kultury* (Studies in the Theory of Culture), Warsaw 1956, p. 65.

Jerzy Kmita/Poznań

THE IDEA OF ANTAGONISM OF ART AND SCIENCE
AS A *WELTANSCHAUUNG* COMPONENT

1. The antagonism of art and science

Within philosophy, the problem of a mutual relation between art and science usually assumes an evaluative, axiological character. What it means is, by and large, that, when formulated, this problem is meant to reflect one's attitudes towards the values which, on the one hand, are connected (in a given case) with art and, on the other hand, with science. However, these two disciplines can also be compared in a different way. First, it can be done through psychological investigations; we are then interested in such questions as the similarities and differences between mental activities comprising the practicing of either art or science, the sorts of skills or predispositions necessary to the practicing of the above disciplines (they either promote or hinder that practice), the course of the process of understanding scientific utterances or statements and receiving art productions, etc. Incidentally, the psychological investigations which have been carried out so far clearly indicate that various scientific and artistic disciplines enter into relations with all kinds of intellectual operations or skills, which makes, e.g. the practicing of mathematics be relatively the closest, psychologically that is, to writing poetry or composing music, while the empirical natural sciences are in this respect corresponding to realism in literature or painting. Secondly, the comparison that we are now considering can also be done on the level of the theory of culture. In this case, we shall be comparing values commonly respected within particular communities of scientists and representatives of art; we shall also be comparing the methods of their realization commonly respected within those communities. And thirdly, taking the above level into account, we may assume a sociological viewpoint and concentrate upon the similarities and differences between sociological determinants and functions of the two disciplines of human activity.

In this paper, however, we shall be dealing mostly with the philosophical aspect of our problem and we shall only mention its sociological and cultural context.

And so, in a philosophical context, this problem is usually solved in the spirit of the thesis of antagonism which propagates the existence of a basic conflict juxtaposing the two disciplines in question. This thesis is, at the same time, formulated in the way oriented axiologically, placing its representative on one of the sides of the conflict. As we have already mentioned, it seems to be a typical situation in philosophy. If we are dealing with the preference of scientific values, we shall speak of scientism; in the other case — we shall speak of anti-scientism. Let us observe that, although a philosophical articulation of the conflict between science and art has its ancient antecedents (beginning with Plato, who ruled out poets from his ideal world), this conflict originates at the moment of forming of science, i.e. in modern times.

The thesis of antagonism is usually connected with the following ideas:

1. Between the fundamental tasks (aims) of a scientific and artistic activity there occurs a significant difference as well as a substantial axiological conflict.

2. The fundamental tasks (aims) of the scientific and artistic activity are, as a rule, of the same type. In both cases, the value in question is cognition; however, the respective varieties of cognition are basically antagonistic; one of them is, at the same time, at least "inferior" to the other one.

3. There is no direct axiological conflict between the aims of the scientific and artistic practice. However, the aim of science is "inhuman" — which is an idea that represents only the anti-scientific tendency. Science observes no "human sense" which means that this sense is not connected with the "experienced social world" *(Lebenswelt)*, i.e., with the common experience and the worldview ("*Weltanschauung*") that fills it.

4. The sense of science and art is of a cognitive type and the conflict between these disciplines of human practice is caused by the diversity of methods used within these disciplines in order to realize that sense. In fact, it is a conflict between "logic" and "intuition". It is usually considered together with the conflict discussed in paragraph 2.

The sense of the above ideas, as it has already been stressed, is not that of cultural descriptions but rather that of a worldview. The reason for that is that they constitute the components of certain worldviews (here: the scientific or anti-scientific one), i.e., such systems of beliefs which create a complete image of the world, characterized by the (explicit or implicit) presence of superior values in it as well as by properly relating the practically perceivable values — "life" — with these superior values ("the sense of life"). This relation either makes particular aspects of "life" positive values, leading towards the realization of superior values, or, at least, favouring this realization, or it makes them negative values which obstacle this realization and sometimes make it impossible. Certain aspects of "life" may stay beyond this kind of worldview valorization. Referring to a certain terminological tradition that appears in modern philosophy, we shall call the socially functioning "bundles" of ideas, included in a given worldview and characterized by (1) the fact that, at least implicitly, they have an evaluative (normative) sense, (2) the fact that they do not imply any technologically effective (in a broad sense) directive judgements, myths.

Two comments seem to be necessary at this point. And so, first, the technologically effective directive judgement is the one that adequately points out to a certain activity as to a condition (in certain practically perceivable circumstances) sufficient or/and necessary to obtain such and such a practically perceivable effect. This effect may be of a highly diversified nature (this being a reason for our mentioning the *broad* sense of the technological effectiveness); it may be a craftsman's, industrial, agricultural, economical, medical, pedagogical, the so-called sociotechnical, communicational (linguistically communicational, customary communicational, and even artistically communicational), etc., effect. Directive judgements may, at the same time, be verbalized as directives (thus the name) of the following form: "In order to obtain E, it is necessary (in circumstances C) to undertake the activity A", although they may also be expressed in the form of descriptional conditional statements. Second, in the meaning suggested here, a myth should not be understood either in the classical ethnological sense (a story peculiar for the so-called primitive cultures), or in the common sense (a kind of fiction or even untruth told purposely). This word must be understood exactly in the way it has been described above: it may signify a story (with plots) or an abstract theological or philosophical treatise. Moreover, referring to a myth, we shall not use here any epistemological qualifications such as truthfulness/falseness, validity/invalidity, legitimacy/illegitimacy, etc.

The above decision requires a certain argumentation.

Let us begin with a declaration that we shall understand the scientific practice (science — in one of its meanings) here as a methodologically regulated activity, the product of which are certain statements or theories characterized by a direct or (mostly) indirect technological effectiveness. That means that, using as a rule certain additional premises, we may conclude from these statements or theories (in a more or less direct way) certain directive judgements, directives which are potentially effective technologically. These statements or theories are obtained (or respected as theories in a scientific community, to be more precise) with the use of certain methodological norms and directives, of which people are not clearly aware. The norms and directives constitute the social methodological consciousness of a given scientific discipline at a given historical period. These norms establish such cognitive values as truth, validity, legitimacy, etc., with the assumption that the fundamental value is constituted by truth understood as conformability with the "objective reality". And so, it is easy to notice that the statement including the so understood truth is not characterized by the technological effectiveness (since the "objective reality" is not practically perceivable) which leads to a conclusion that the concept of that truth is a myth. It is a scientific myth, too, because it is included in the worldview which exposes this truth as a central value. And yet, it is possible to use these normative epistemological qualifications, included in the scientific myth, abstracting from the latter circumstance. It is possible because these qualifications conceal a certain non-mythical content: the fact that respective statements or a certain theory have been obtained according to the methodological directives specified for the given discipline, at a given point of its historical development. This fact provides this qualification with a certain non-mythical sense. On the other hand, if we refer these qualifications to the myth, we cannot expect them to have this sense whatsoever. We can only consider them from the point of view of the sense of the axiological approval which is too flimsy to base on it the intersubjective communication. For that reason, it is necessary to give up the epistemological qualifications of a myth and reserve them exclusively for those judgements that are characterized by the (mostly) indirect technological effectiveness (i.e. the judgements obtained in science).

And so, as a consequence of that, the thesis of antagonism between art and science should be considered as a myth which can be essentially approached only with the indefinitiveness of the intersubjective declaration that expresses this approach. However, it is also possible to take a scientific approach to this myth. Taking such an approach requires only to ask about its determinants and social functions. We shall now take the latter approach, regardless of the fact whether it is the scientific or anti-scientific version of the myth of antagonism that we are dealing with.

2. The anti-scientific version of the myth of antagonism between art and science

As a matter of fact, until romanticism in modern European culture, and most of all in philosophy, the most exposed idea is the one according to which art and science are complementary as long as their objectives and methods are different. This idea opposes another one, according to which they remain in a conflict. Only in the era of romanticism has the myth of antagonism developed. As a component of the scientific worldview, it expressed a generally forbearing disrespect for "artistic phantasies", while in its anti-scientific version, it assumed the form of a fierce attack upon the "rationality" of modern natural scientists and their philosophical advocates (who, very often, were the natural scientists themselves — as the representatives of the "philosophy of nature", the name used at that time for mathematical natural sciences). This attack is connected with an intense apology of the artistic "sensibility and faith", as well as with recognizing at least the ontological equivalence of the world cognizable through that "sensibility and faith" and, at the same time, its ethical superiority (not even mentioning the aesthetical one).

The Enlightment worldview, which considered the order of "Nature" as a basis and, at the same time, the subject of explorations of scientific natural sciences, has, at the same time, considered this order as a criterion of the artistically-aesthetic qualities of art. From an "ideological" point of view of values, it liberated art from its feudal religious and courtly servitude. This worldview, however, which has emancipated the artistic practice so effectively for some time, started to be rejected by the same artistic practice as the order of "Nature" was more and more clearly becoming the technological basis for a new, capitalist organization of the social world, while the worldview itself was becoming the apology of that world, the apology of (technological) "progress" towards which it was supposed to be tending. The Enlightment worldview has lost, what has been called by the representatives of the Frankfurt School, its "critical" (antifeudal) functions in relation to the current social reality and the myths that made it legitimate. The scientific practice of the natural sciences has thrust aside art, which — being a type of social activity — participated only to a small degree in the stabilization of that reality and, for that reason, became the subject of a disrespectful forbearance.

It is obvious that the social classes which, so far, have all been using the ideal of the education of Enlightment, revealing to the society the order of "Nature" and criticizing from that point of view the current social reality, have been antagonized by a new, capitalist system of production and the presently emerging legal and political system, sanctioned by that order and adjusted to the new system of production. It would, however, be an oversimplification to consider the anti-scientific mutiny of the romantic art exclusively as an expression of the class protest against the new capitalist order of the world. It is a striking fact, for example, that within romanticism, both the pro-feudal conservative ideas and the plebeian anticapitalist ideas are known to have been mixed together. The point is that for the artistic communication of an aesthetic image of the world, a grave danger is its scientific image. The reason for that is this communication requires a social vitality of myths eliminated by the scientific image of the objective (metonymical) relationships between the practically perceivable, technically exploited phenomena. The leftover is the scientific myth, which is too indigent for the purposes of artistic communication (let us observe that even the later "positivist" literature has not stopped at this point). And so, it finally appeared that the

"Nature" of natural sciences and the "Nature" of art are totally contradictory to one another. The former, as it has been said by Goethe, appeared to the advocates for art as "nature given away to tortures".

And so, the anti-scientific myth of antagonism is functioning socially not only as the negation of the new organization of the social world and its scientific-technological base, but also as an apology or an attempt to revive the "art feeding" myths taken either from the medieval religious and knightly tradition or from common beliefs and legends. In its defence of the semantic fundamentals of the artistic practice, the romantic worldview also makes use of the contemporary philosophical thought, particularly the idea of German classical idealism: the domination of the Spirit over the world of the mathematical-experimental "matter" — the "tortured" Nature considered by Hegel to be only a mediated Spirit. However, it has particularly been the intuitionism of Schelling that was suitable to support the view concerning the superiority uniting the spirit and the nature, i.e., the form of aesthetic cognition over its competitor in the form of the truth of the scientific rationalism.

Let us observe that, while the romantic anti-scientific myth is truly involved in the fight for the "life" of art, its scientific counterpart struggles only for the worldview valorization of the scientific practice. As the latter develops, the need for that valorization decreases so much that the scientific myth looses its popularity within the practice of natural sciences. Beginning with the end of the 19th century (conventionalism), the ideal of reflecting the "objective reality" gradually yields to such values as a peculiar aesthetics of formal simplicity, intellectual economy, mathematical elegance, etc. On the one hand, the mathematical natural sciences have gone so much astray from its original pretheoretical form, closely relating it with common practical wisdom, that there is no need any more to contrast its practice with other types of social practice. On the contrary, the chains binding the naive scientific methodology appeared to be too tight thus making the "aesthetic" myths or the myths of more exquisite forms of scientism more functional. On the other hand, the common practical wisdom mentioned above, ceased to be the "bone of contention" with the artistic practice, for which it constitutes, as much as it is realistic, the basis of the communicated image of the world. At the moment at which the theoretical natural sciences loose their relation with the common practical wisdom, the *Lebenswelt*, as it has been adequately observed by Habermas, the scientific defence of one's distinct character becomes unnecessary. Natural scientists and their philosophical advocates do not oppose "artists" at all any more. On the contrary, they start considering themselves as ones.

This sympathy, however, is not reciprocated by the representatives of art, which is partly caused by the increasing aggresiveness of scientism

59

connected not with the theoretical natural sciences but with the Humanities (we shall speak about that later). However, it is mostly caused by the fact that the image of the world produced by the "artistic" natural sciences still does not tolerate myths, moreover, it becomes "inhuman" in the sense already discussed. What is even more, the more and more effective (technologically) scientific knowledge turns the world into the hands of technocrats, thus making it a desert for art, a desert on which rules the mass culture rejecting art in favor of the entertainment industry.

It is, at the same time, characteristic that this mass culture adopts the most naive forms of the scientific cult of science, associating them with the relish for parascientific beliefs, so strongly condemned by the scientism of philosophers. And yet, these beliefs, accepted because they are taken as a product of science, cannot become a basis for a new vitality of art. The reason is that the values, the realization of which they secure, do not reach beyond the circle of values of the consumptionistic worldview, characteristic of mass culture and not leaving any room for artistic creativity in which there occurs a peculiar unification between "life" and the nonpractical, "metaphysical" values symbolized by that life's various aspects. The so-called "science-fiction" often connected with such beliefs must, similarly, be considered as part of the entertainment industry rather than as part of art.

3. The scientific version of the myth of antagonism between art and science

The factors which caused that, in the past, the scientific myth of the antagonism between art and science was somehow functional in relation to the scientific practice of natural sciences, are also significant for its functionality in relation to the scientific practice of the humanities. This, however, is true only to such a degree to which it actually is scientific, i.e., to which it produces the indirectly effective (technologically) statements and theories.

On the one hand, the degree of the so understood scientific character of the humanities is not high. This should make us think that the mentioned myth cannot experience any particular popularity within the academic humanities. On the other hand, however, the fact that the humanistic disciplines, in their scientific aspect, usually represent the pretheoretical level, which has been long ago and to a different degree exceeded by the theoretical natural sciences, and thus they are closely related with the common practical wisdom and they represent its cognitive horizon, should, on the contrary, make us predict the vitality of the scientific myth of antagonism in the humanities. This vitality would be a result of the already mentioned "bone of contention" in the form of the common practical wisdom utilized together with art. Moreover, the range of that wisdom is almost identical for both art and the humanities, which is contrary to the natural sciences.

And so, we are dealing here with a situation in which there are two, opposed to each other, tendencies. One of them is created by the desire of the humanities to preserve the feeling of their distinct character in relation to the artistic practice, which would make the scientific myth of antagonism functional. The other one is created by the fact that the technological, science-creating function is not of a primary importance to the practice of the humanities, which highly reduces this desire. And yet, this reduction is not, in any case, explicitly determined. The reason for that is that the non-technological social functions of the humanities — the myth creating ones, the ones of cultural education and worldview valorization of "life" — overlap, to a certain degree, with the social functions of the artistic practice, which, in turn, brings about a peculiar competitive-professional rivalry.

And so, by and large, the attempt to position both the humanities and art together is full of ambivalence. It has also become the reason for the fact that, among the representatives of the academic humanities, one of our myths has never become a dominating one. On the contrary, they have usually coexisted, fighting one another fiercely only in their philosophical verbalization.

For example, the terminological tradition which puts the humanities into the range of science is of German origin *(Wissenschaft – Geisteswissenschaften* or *Kulturwissenschaften)* and has been popularized in Slavic countries, while the French or English term "science" has, until now, meant natural sciences, even though it is presently also used in such contexts as "behavioral sciences", "social sciences" or "cultural sciences" which deal with certain branches of the humanities. However, it has been characteristic of the German antinaturalistic philosophy of the humanities of the turn of the 19th century, which considered *Geisteswissenschaften* or *Kulturwissenschaften* as undoubtedly a scientific discipline, to view this discipline as being related with art but, on the other hand, to oppose it to the natural sciences. And it is also beyond any doubt that it does so from the scientific position. The paradoxicality of this state of affairs is even greater if we consider the fact that, at the same time, in the countries, the terminological tradition of which excludes the humanities from the scope of science, it is the former that is practised according to the model of the anti-scientific positivism (Mill, Spencer, Buckle, Renan, Taine or Durkheim). And so, even if the desire for an indirectly technologically effective humanistic knowledge has determined the vitality of the scientific myth of antagonism, it was not in Germany of the end of the 19th century. One should rather believe that the ennoblement of the myth-creating and educational function of the humanities that was carried out there with the help of the word "science", repleted with definitely positive axiological content, has simply functioned in the professional interest of a large group of university scholars (let us have in mind that universities have been the main institutional base of the scientific life in Germany).

It must also be stressed that this philosophical, antinaturalistic articulation of the anti-scientific myth of art and the humanities, turning the humanists into second-rate "artists" who persuade culture and its myths much less effectively than the "true artists", has not, so far, had any significant effect upon their practice. Commonly respected is the type of a peculiar, not much aggressive in its scientism, positivistic worldview with an exposed devotion to the "professionalism" (the mechanism of footnotes). Probably only the representatives of particular art sciences are less consistent in this respect, trying to win their popularity within artistic élite by way of special coquetry.

The scientific myth of antagonism has been most energetically exposed by the neopositivistic philosophers, the representatives of the so-called logical empiricism. Although the methodological program has been formulated by that orientation mostly for the purposes of natural sciences, it has, nevertheless, been so much historically delayed that it had a practical influence only upon the humanities and particularly upon that trend of it which, in the form of "behavioral sciences" or "social sciences", tried to function as a source of "verifiable", technologically effective knowledge. The stress put upon the precision of language, which was supposed to oppose science to the "irresponsible artistic mumble", can be explained not only by the fact that giving precise definitions of the concepts of common practical wisdom and a logical arrangement of knowledge accumulated within the scope of that wisdom is always characteristic of the early, pretheoretical stage of development of any branch of the scientific practice, i.e., the practice of producing the indirectly technologically effective knowledge. This general regularity is accompanied by the fact that the competition between the artistic activity and the academic humanities occurs not so much in respect of the myth-creating function but rather in respect of the persuasive communication of both the myths and the elements of common experience. The reality shown in art masterpieces is a more effective device of that persuasion than the generally discursive humanistic utterances inasmuch as it presents the relevant contents in a somewhat concealed form. The communicated aesthetic image of the world is "embedded" in that reality and constitutes its implied meaning. It is usually assimilated without the recipient being aware of it and spontaneously set by him in motion as an implicitly functioning set of assumptions of the semantics of artistic expression which concentrates his attention upon the presented reality itself. This kind of artistic communication, persuasively effective but not aimed at transferring the (indirectly) technologically effective knowledge, must be opposed by the self-conscious science to the language accuracy which does not tolerate any artistic "tricks" that introduce ambiguities so unfavorable for the technological effectiveness of the communicated knowledge.

Besides neopositivism, the other intellectual orientation which exposes towards itself the scientific myth of antagonism in a similar way is structuralism. It seems to be an interesting orientation inasmuch as it not only influenced the humanistic scientific practice but also, which is paradoxical, functioned as a program of artistic activity, particularly in the version called *Russian formalism*. Let us now scrutinize the historical coincidence which caused that the structuralistic scientific ideas were assimilated by art and served its practice. It is not, in any case, an exceptional situation in history; it is enough to mention the literature of the French naturalism or the Polish positivism. And yet, never before has the artistic creativity scientifically based been so radically opposed to any form of art devoid of scientific grounds.

Structuralism has been known to have been formed first as a certain theoretical concept of linguistics and a closely related to it methodological program of investigations from that range. In its de Saussurian form, it has inherited the scientific orientation of the positivist early-grammatical linguistics, supported by the intellectual and equally scientific Durkheimian inspiration. It is worth mentioning here that linguistics has always been, generally speaking, more invulnerable to the anti-scientific ideas expressed with particular emphasis by the German philosophy of the humanities than the remaining *Geisteswissenschaften*. Even the later Crocean conception of linguistics as one of the disciplines of the aesthetics of expression, as well as the neoidealistic conceptions, have not managed to maintain for long their position as alternative to structuralism investigative programs (mostly in respect of the stylistics of literature).

This quite emphatic, scientific tendency of linguistics can probably be explained in the following way: firstly, the cultural-educational function of that discipline (the initiation into respecting the norms and rules of a language) does not enter into any significant conflict with its precisely scientific function (providing the technologically effective premises for linguistic communication) because the pretheoretical registration of the rules or premises of a "correct speech" constitutes, at the same time, the device used for their cultural and educational persuasion. Secondly, the myth-creating function of linguistics, which destroys potentially its scientific character, is, in fact, quite restricted. The only myths we can include here are the myth of a general humanistic spirit (derived from a homogeneous biological set of properties of particular individuals), the myth of a national spirit and the myth of their development, followed through the historical evolution of phonetics, grammar and, most of all, vocabulary (etymology).

Structuralism has offered a methodology which significantly increased the degree of functionality of both strictly scientific and cultural-educational linguistics, limiting, at the same time, its myth-creating capabilities. Therefore, the popularization of that tendency among linguists must be considered on the one hand, as an expression of a relevant development of linguistics towards its more "scientific" form and, on the other hand, as an expression of science-creating demands influencing this discipline. This influence yields an explanation of its scientific orientation, which is much stronger than in the case of early grammatical linguistics.

For a specialist in the history of linguistics, the above statement is beyond any doubt. In order, however, to make it much more generally obvious, let us observe that the system-like structuralistic antipsychologism — considering particular language regularities (phonetic, syntactic and semantic ones) not as an effect of mental associational processes but as an indication of system constituting paradigmatic (alternative oppositions selected for uttering language units) and syntagmatic (linguistically authorized arrangements constituting sequences — utterances) relations — wins an opportunity for a much more adequate registration of language norms and rules. These rules and norms are not connected with our experiences, but they define the methods of manipulating the language units. Moreover, these methods of manipulation coexist and constitute a system on the synchronic level. A separate, historical treatment of every phraseological unit cannot ever lead to obtaining an adequate reconstruction of the synchronic system of this type.

And so, this is the way in which the scientific orientation of the linguistic structuralism should be explained. At the same time, within the range of artistic practice, we are dealing with a modernistic attempt to instaurate the myth of an artist as the Nietzschean *Übermensch* which, as a myth of an individual provided with a "creative power", constitutes in art — able to function socially and communicate its aesthetic world visions only on the basis of socially respected myths — the only counterproposition to the socially dying out myths of a scientific civilizational "progress" and the "metaphysical" myths contrary to it. The myth of an artist — the spiritual aristocrat — is unexpectedly gaining a much newer form — the one of an artist who participates in the revolutional labor movement and, together with this movement, bases its activity on scientific grounds. Being a proletarian (an engineer) working with a pen or a chisel, such an artist, just as a proletarian or an engineer using in his production practice certain "scientifically" established objective regularities, needs a similar "scientific groundwork". And he finds such a groundwork in a formalistic extrapolation of structuralistic language studies over art phenomena. Such an extrapolation has been the Russian formalism. Connected with it is the myth of an artist who is not any more a revelator of truths inaccesible to science. Breaking away, in a totally earthly way, with the mysticism of the art of the past, he uses the "scientifically" investigated regularities of an artistic system, constituting the equivalent of the regularities of a linguistic system. At the same time, he uses those regularities by implementing certain "artistic manners".

And so now, the content of this peculiar implementation of the scientific myth of the antagonism between science and art in the artistic practice becomes clear. This antagonism, which can be presented in the form of an opposition between science and art scientifically grounded, on the one hand, and art void of such grounds, on the other hand, is — in fact — a manifestation of the juxtaposition between the traditional, "mystical" bourgeois art and the "scientific" new proletarian art.

Despite the fact that the myth of an artist rebellious against the traditions of the bourgeois culture of a proletarian (that working with a pen as well as that working with a chisel) has been forgotten first, which seems paradoxical, wherever the revolutionary aims of the labor movement have been institutionally taken over by the state which originated from that movement, structuralism — as a type of humanistic studies in language and art — has not lost its scientific orientation. What is even more, it has preserved it as a general theoretical-methodological conception of studies in culture, understood as the studies in social communication (i.e., the economical, moral, artistic and linguistic one). It is equally represented by the structural anthropology of C. Levi-Strauss, who argues against the anti-scientifically minded existentialism, and the theory of culture adopted by the so-called Tartu school. And yet, the continuators of Levi-Straussean structuralism have, in fact, given up the ideals of scientism; also, we will not find in them any trace of the myth of antagonism between art and science. No wonder that C. Levi-Strauss separates himself from contemporary French structuralists; he says he simply does not understand their ideas.

The formalistic and structuralistic as well as neopositivistic version of the myth of antagonism between art and science, representing the scientific worldview of the humanistic and even artistic practice, have been the most recent creative articulations of that myth so far. They are still supported in various epigonic statements but the modern creative initiative belongs to their anti-scientific opposition. Drawing its main inspirations from the philosophy of Heidegger and its particular developments, this opposition declaratively denies any intention to undertake the question of the antagonism between art and science. However, in fact, it takes a well-defined position in this respect. It takes this position, thus depreciating the scientific ("mathematical") thinking and persuading an attitude that cannot be reconciled with the behavior to accumulate the technologically effective knowledge.

Translated by Krzysztof Sawala

LITERATURE

1. Condorcet A.N., "On Future Developments of Human Spirit", (in:) A.N. Condorcet, *An Outline of the Picture of the Development of Human Spirit Throughout the Ages* (Polish translation), Warsaw (1957).
2. Dilthey W., "Poet's Imagination. The Elements of Poetry", (in:) W. Dilthey, *Aesthetic Writings* (Polish translation), Warsaw (1982).
3. Dilthey W., "Novalis", (in:) W. Dilthey, *Aesthetic Writings* (Polish translation), Warsaw (1982).
4. Feyerabend P.K., "On the Critique of Scientific Reason", (in:) *Essays in Memory of Imre Lakatos* (ed. R.S. Cohen), Dordrecht (1976).
5. Habermas J., "Technology and Science as 'Ideology'" (Polish translation), (in:) *Sociology Crisis?* (ed. J. Szacki), Warsaw (1977).
6. Grzegorczyk A., *Nomothetical Models of the Development of Literature* (in Polish), Warsaw/Poznań (1983).
7. "The Sorrow of the West. An Interview with C. Levi-Strauss", (the Polish translation) *Prezentacje*, 1982, no. 3.
8. Zeidler A., *The Anti-naturalistic Program of Studies on Art and its Modern Continuations* (in Polish), Warsaw/Poznań (1983).

II. IDEOLOGY OF THEORIES

Jerzy Kmita/Poznań

SCIENTISM AND ANTI-SCIENTISM

This paper is an attempt to outline a certain interpretation of the thesis concerning the realistic character of the scientific (most of all) cognition. As it will be shown in the conclusions, I have termed the point of view which accepts both the interpretation and the thesis itself as meta-scientism. I have decided upon that name because, among other things, it is directly a standpoint towards the two opposed trends of philosophical thinking — the scientism and the anti-scientism. A critical analysis of these trends has been the basis for the present considerations.

I. THE NAIVE VERSIONS OF SCIENTISM AND ANTI-SCIENTISM

1. Three Constituent Views of Scientism

This does not seem to be the best place to involve oneself in any analytical operations demonstrating a significant divergence between the possible uses of the terms: scientism and anti-scientism. It would undoubtedly constitute a separate assignment for the otherwise significant detailed historical investigations revealing the variability of the ways of functioning of the orientations called in the above way as well as the variability of the values exposed by them. Adopting the below given meanings of the terms, I am prompted by two reasons. Firstly, it seems intuitively obvious that a majority of modes of thinking called "scientific" or — on the contrary — "anti-scientific" can be located, according to that terminology, respectively on the two sides of the dychotomy suggested here. Secondly, the criteria that I have used in order to establish this dychotomy seem to constitute a convenient starting point for grasping certain important differences separating the traditional forms of scientism and anti-scientism from the modern ones. Incidentally, I have differentiated between these two forms with the use of terminology borrowed from Lakatos, although used it in a different context. And so, I have called the traditional form of scientism or anti-scientism the naive form, while the modern one has been called by me the sophisticated one.

I have assumed that the following three beliefs are constitutive for scientism:

(1) The following two relations between a judgement concerning a given state of affairs and the state of affairs itself are possible: either the judgement corresponds with the state of affairs (it is adequate) and then we say it is true, or it does not correspond with it (it is inadequate) and then we say it is false.

(2) The acquisition of true judgements (true in the above sense), or (positive) conceptual cognition (since judgements are of a conceptual character), is the only possible form of cognition.

(3) Conceptual cognition is a fundamental value.

I shall first make a brief comment on statement (3); then, I shall procede to the more complicated problems resulting from statement (1). Statement (2) seems to require no comment whatsoever.

The "fundamental value" is understood here as either (a) the "ultimate" value, i.e., the one not subordinated instrumentally to any other value, or (b) the value which constitutes the necessary instrumental means of realization of the "ultimate" value ("ultimate values"). Therefore, it may happen so that an advocate of scientism considers the acquisition of conceptual truth as the "ultimate" value ("the value in itself"), or he may distinguish, for example, progress as the "ultimate" value, instrumentally superior to the acquisition of truth. In the latter case, he considers conceptual cognition (the acquisition of truth) as necessary means of the self-realization of progress.

I would like to make the following comment on statement (1): firstly, it talks about a judgement "concerning" a given state of affairs and not, for example, "referring" to it. This, by no means, is a coincidence. By means of differentiating between the meanings of these two terms, I am using them to distinguish between the semantic and the epistemological relation between a judgement (a statement) and a given state of affairs. I have expressed the semantic relation in the form of the formula saying that referent of a particular judgement (statement) is such and such a state of affairs, while the other, epistemological relation has been expressed by me with the help of the word "concerning". What is the difference, now, between these two, usually not distinguished from one another, relations? The essential difference lies in the fact that, while a given state of affairs, fixed by a judgement (statement) which "concerns" it, is determined as the real one, the state of affairs which is a semantic referent of this judgement (statement) is not, in advance, defined in terms of the opposition: real — unreal. The semantic referent of a judgement expressed by the statement: "Warsaw lies on the Odra River" is the following state of affairs: the fact that Warsaw lies on the Odra River. At the same time, this judgement might concern, among other things, such states of affairs as the fact that Warsaw lies on the Vistula River, the fact that Warsaw does not lie on the Odra River. etc. What is important is the "scientistic" ways of characterizing the epistemological relation to which the term "concerning" corresponds. None of them, as we shall be able to see, can be accepted in the epistemological perspective assumed by me. Equally unacceptable within this perspective is the conception of the comparability between judgements and certain real, single states of affairs. For that reason, statement (1) should, as a matter of fact, be interpreted as if it were in quotation-marks.

Certain features are shared by what I have called a semantic referent of a judgement (statement) and a phenomenological intentional state of affairs. There are, however, certain significant differences here. They appear as, for example, the fact that while the intentional state of affairs must be given to the consciousness and "experienced", the semantic referent "originates" elsewhere, in the sense that it does not have to be "directly given to senses or intellect". And so, the best thing would be to forget this comparison, so that it does not lead us to a mistake.

A more precise explanation of how, in my opinion, the semantic referent of a judgement (statement) originates, would require a separate investigation; nevertheless, I have to say a few words about it since this concept will soon be necessary for our further considerations.

Semantic referents are related to particular statements or judgements expressed by them through a set of specific directive judgements, commonly respected in a given community, which will be called here semantic rules. This term is somewhat inexact because these rules are, as a matter of fact, only verbalizations of these semantic directive judgements. Having mentioned this inaccuracy, we may now try to imagine these rules as denotation rules (with a generalization that also the logical constants may denote). The states of affairs, being the semantic referents of judgements (statements), are set by the commonly respected denotation rules with a precision similar to that with which these rules are commonly respected (but not necessarily consciously applied). And so, I consider both the rules of denotation and the semantic referents of linguistic utterances, particularly statements (judgements), as substantially social. This does not mean, however, that it is impossible to consider the individual variants of these rules of reference as well as the individual variants of the semantic referents which correspond to particular judgements (statements), in the form of specified states of affairs. The formation of these individual variants is determined by certain psychological mechanisms, which act, above all, within the process of the practical acquisition of the social, semantic rules of reference by particular individuals. These individuals use different, characteristic only for themselves, methods of recognizing (a) the elements of particular denotations, (b) particular states of affairs as referents of judgements (statements). These methods include, beyond any doubt, perceptive experiences, which are individualized enough to cause the desired differentiation of the semantics of particular individuals. The limits of singular individualizations depend upon, above all, the constraint of interindividual communication in the situations in which such a communication determines the matters that are rudimentary for every individual, like, for example, the universally rudimentary problem of preserving life.

Secondly, to make another comment on statement (1), I must say that the meaning of "real" as applied in advance to a given state of affairs concerned in a given judgement, looks differently in the light of realistically oriented scientism, and differently in the light of idealistically oriented scientism. In the case of the former, the qualification: "real" (sometimes it is called "objective") implies the following statement: the one, for the existence of which it is not necessary that it is ascertained by any judgement. In the case of the latter, "real" means: ascertained by an appropriately qualified judgement.

The idea that an advocate of idealistically oriented scientism relates the predicative word "real" to an appropriately distinguished by himself subclass of semantic referents of judgements (also appropriately distinguished) automatically comes to mind. But even though we might say that "there is something in it", this idea cannot be accepted. Let us first notice that such a person does not use this kind of conception of a semantic reference, a small fragment of which I have just outlined. And secondly, what is even more significant, there are different variants of idealism; only some of these variants would correspond to the opinion that only the appropriately qualified semantic referents of judgements are "real". The variations of idealism that would be of some significance here would be those of the so-called "objective" idealism (the term "objective" being a little misleading). Particularly Neo-Kantianism seems to fit such an interpretation: the real are those states of affairs which are ascertained by judgements (constituting something like semantic referents of these judgements) validated by the appropriate transcendental norms. On the other hand, the socio-semantic interpretation of the epistemological relation manifested in the form of the term "concerning" seems to bear no correspondence whatsoever with the so-called "subjective" idealism. The reason is that this relation is understood psychologically and, thus, also individually here. The state of affairs concerned in the given judgement is a certain individually experienced phenomenon, psychologically determining the passing of that judgement. The adequacy and the validity of that judgement are causally dependent upon the type and circumstances of the respective experiences.

Thirdly, to make yet another comment on statement (1), a belief has been spread that an advocate of scientism always assumes the psychological theory of cognition, which means that he understands the relation of "concerning" as the one occuring between a given judgement and a certain real state of affairs of a physical or mental character. It occurs when the set of perceptive or reflexive experiences elicited by that state of affairs somehow "provokes" the passing of that judgement. Indeed, it seems to be the most typical — for certain traditional variations of scientism — conception of the relation of "concerning", which, in any case, is dominating within the positivistic scientism. However, it is definitely not the only possible conception of scientism. It is accepted, as it has already been said, by the positivistic scientism; it can easily be adjusted within that framework to the "subjectively" and idealistically oriented scientism (the physical states of affairs considered by the psycho-physiologically determined judgements are also, "in the last resort", sets of perceptions), although the "objectively" and idealistically oriented scientism excludes the psycho-physiological theory of cognition. To resort to the most representative examples in this respect, this theory is excluded by the Badenian and Marburgian variations of Neo-Kantian-ism or phenomenology.

Fourthly (the fourth comment), a belief has been spread that an advocate of scientism considers the natural sciences, and particularly the mathematized natural sciences, as the model ones in respect of the method of realization of the basic value, which is cognition. Indeed, in the overwhelming number of instances, an advocate of scientism as-sumes, at least, that this domain is exceptionally effective in yielding true judgements. However, he does not have to claim at the same time that the humanities should adopt the models of scientific procedures used in the natural sciences (it is not done by a "Marburgian" — E. Cassirer, for example). The acceptance of that opinion cannot be withheld either on the basis of the idealistic orientation of scientism, or by its departure from its earlier, naïve form (I shall explain the meaning of that term soon). Moreover, the term "scientism" itself has been created in the countries in which *science* refers to the natural sciences. Despite that, however, the way I have just defined scientism does not determine that an advocate of scientism has to support the opinion analyzed here. E. Husserl, for example, whose beliefs stay within the framework establish-ed by statements (1) — (2), definitely does not accept such an opinion, which is best exemplified in his famous *"Krisis der europäischen Wissen-schaften"*; theoretical natural sciences barely seem to provide for ade-quate (objective) cognition; in fact, such a cognition is a result of a certain deformation caused by being subordinated to their "technological interest", as the representatives of the Frankfurt School would say. The judgements of these sciences select in a biassed, "selfish" way only those elements and relations between the given states of affairs which are technologically useful.

Naturally, it is possible to accept the association of the term "scientism" with the English or French name *science* as valid and assume that procedures used in the natural sciences are the model ones. The above belief can be treated as the fourth constitutive feature of scientism. But the so established meaning of the term scientism will to be too narrow in comparison with its presently appearing usages. The reasons for which I would support respecting these usages would be the following: (a) they make the content of the concept of scientism (and respectively — anti-scientism) more perspicuous and homogeneous and, at the same time, (b) the opposition of the standpoints of scientism and anti-scientism approached to those usages expresses a controversy much more significant for the contemporary philosophical thought than the dispute over whether the humanities are patterned (should be patterned) after the natural sciences. And so, I shall abide by the appropriately established terminological convention: (1) I shall uphold the broader sense of the concept of scientism, (2) I shall, at the same time, ascertain that in the past, closely connected with this orientation was the belief of a model character of the natural sciences in respect of the means used and cognitive effects obtained.

2. Anti-scientism and Methodological Naturalism

It is now necessary to consider the relation between scientism, defined in the above way, and the standpoint called, after Popper, methodological naturalism. Methodological naturalism, the name being slightly inappropriate, propagates (or postulates) the identical character of basic methods of the investigative procedures used in the natural sciences and in the humanities. It also concerns the relation between anti-scientism and methodological anti-naturalism (the negation of methodological naturalism). Before we procede to that, let us first spend some time on anti-scientism.

This orientation could be defined by that it negates at least one of the three constituent beliefs of scientism discussed above. Although there are several (seven, to be precise) theoretical combinations, we shall distinguish here only some of them — those, which occur in reality. And so, an advocate of anti-scientism may: (a) negate statement (3) without negating statements (1) and (2), and particularly without considering them at all; (b) negate statements (2) and (3) without negating statement (1); (c) negate all statements (1) — (3).

It seems that the most popular form of anti-scientism, also outside philosophy, is represented by (b). One may speak here of a romantic anti-scientism because it is characteristic of romanticism. Acknowledging the possibility of obtaining "the truth of reason" — the conceptual cognition, romanticism as a rule opposed to it other values, which it considered as more significant, thus denying it the name of the fundamental value. Romanticism includes among these other values also the cognitive value (different from the conceptual cognition, i.e., "the truth of heart") but as intuitional and irrational cognition — contrary to intuitional and intellectual (the conceptual one), Cartesian or, later, phenomenological cognition. It can be seen that we are dealing here with the case (b) rather than (a). All the advocates of intuitional and irrational cognition are, naturally, advocates of anti-scientism; their anti-scientism, however, does not have to be of type (b), which is romantic and moderate: apart from "the truth of heart", there also exists an inferior to it "truth of reason". Also important seems to be the case (c): the only truth that exists is "the truth of heart".

Before I show that the ranges of concepts of (1) scientism and methodological naturalism, and (2) anti-naturalism and methodological anti-naturalism do not overlap, I would like to make a comment concerning the term "methodological naturalism" (and — respectively — "anti-naturalism").

As it has already been mentioned, the term "methodological naturalism" (and its antithetic counterpart) is slightly inappropriate. The point is that: (1) it can easily be associated with the image of the natural determination of "the humanistic world", culture, while a general standpoint defined with that term does not forejudge such a determination, (2) this term suggests that the uniformity of basic investigative methods in the natural sciences and in the humanities is a result (postulated or real) of the humanities imitation of the natural sciences, while this uniformity can also be a result of something else. For example, it can be a result of acknowledging a humanistic character of the object of investigations of the natural sciences (which may undoubtedly sound shockingly for an adherent of positivistic scientism). And so, in order to avoid the undesired associations, instead of the term "methodological naturalism" (and, respectively, "methodological anti-naturalism"), I shall use the term "methodological unism" ("methodological anti-unism").

That an advocate of scientism does not necessarily have to be, at the same time, an advocate of methodological unism, is quite clear from the example of the philosophy of E. Cassirer. Conceptual cognition, and so — in this case — the one regulated by some kind of a symbolic form, is, for Cassirer, characteristic not only of science, understood here as the mathematized natural sciences, but also of all other domains of culture, such as myth, religion, language, etc. To each of these domains, a different symbolic form is ascribed, which regulates the construction of a reality characteristic of the given domain in a different way. The reflexion on culture, i.e., the humanities, is not regulated by the symbolic form which organizes the construction of the reality of the mathematized natural sciences. For that reason we are dealing here with methodological anti-unism coexisting with scientism. This case is by no means less typical than that of W. Dilthey's standpoint, in which the romantic anti-scientism (the superiority of intuitional-irrational cognition over conceptual cognition) is associated with methodological anti-unism (it is claimed that it is the humanities, contrary to the natural sciences, where the superior, intuitional-irrational cognition takes place). Besides E. Cassirer, we may include here the representatives of the Baden School, phenomenologists, and E. Spranger (naturally, they all provide different arguments for their opposition towards methodological unism).

On the other hand, an example of an anti-scientism advocate who, at the same time, is an advocate of methodological unism, is L. Althusser. According to him, the production of the so-called "object of cognition" by science has nothing to do with the adequacy in relation to the real states of affairs. He calls the opposite standpoint "empiricism" — in the sense of our "scientism". "Theoretical anti-humanism", which he assumes, leads, however, to the standpoint of methodological unism.

3. The Naïve Versions of Scientism and Anti-scientism

Naïve scientism will be understood here as a standpoint which is created by adding a fourth statement, to the three already discussed statements constitutive for any form of scientism. I shall formulate that statement in the following manner: (4) there exists a presuppositionless and originary cognition (a conceptual one, naturally).

It concerns the type of cognition which is included in the phenomenological *Anschauung*. This *Anschauung* may be represented here by experience, in the usual sense of this term, or by other types of "perception". I use the terms "presuppositionlessness" and "originarity" in my own, rather than the phenomenological, sense. And so, that a given cognition is presuppositionless means that the true judgements which represent it are obtained without assuming any other judgements, while they refer to a certain kind of direct confrontation with the states of affairs they concern, or, in other words, to the fact that these states of affairs are given *in Anschauen*. On the other hand, the fact that a given cognition is *originär* means that all the judgements which do not express this cognition must, in order to show their truthfulness, be justified with the use of judgements of the first type. The particular and most common expression of the so understood concept of originarity is a belief that the definite logical relation between the originarily obtained judgements and the other judgements must be revealed. The understanding of the term "must" in this context implies that the necessary condition to show the truthfulness of judgements which are not originarily given is to reveal the appropriate logical relations between them and the originary judgements (which, at the same time, are presuppositionless). However, by and large, this condition cannot be considered sufficient to show the truthfulness of non-originary judgements.

Naïve scientism understood in the above way, i.e., scientism acknowledging the presuppositionless and originary cognition is represented by various philosophical orientations, which, in many respects, are contradictory. I shall ilustrate this thesis with two examples: positivism (together with neopositivism) and phenomenology.

For positivism, the presuppositionless, originary judgements are judgements which are "directly based on experience". Regardless of whether experience is understood as individual experiencing of particular classes of perceptive sensations, or as a result of the psychophysiological perception of external states of affairs, these judgements are fundamental for any cognition. The second understanding of experience stresses a tendency not to treat this fundamental element too extensibly. The point is that, while introspective judgements which register the "internal" states of affairs cannot be false, the judgements concerning the external states of affairs that "provoke" introspective judgements can be false as a result of various disturbances in the perceptive psychophysiological mechanism. This, however, does not deny the presuppositionless and originarily-based character of all the true perceptive judgements. The fact that, passing perceptive judgements concerning the external states of type of circumstances accompanying the passing of these judgements (the lack of disturbances or the presence of disturbances) is concerned, does not mean that displaying the truthfulness of the judgements of this type always requires an additional acceptance of further judgements concerning the circumstances which accompany the passing of the former (in this case, a direct empirical cognition would not be presuppositionless). There exist certain perceptive judgements, the truthfulness of which is beyond any doubt, and it is they that constitute the foundation of our entire cognition.

To be more precise, on that foundation rest judgements which correspond either to all kinds of statements formulated within science (Millean radical empiricism), or to all kinds of non-analytical statements (the Humean-neopositivistic tradition), or to all kinds of non-analytical and non-theoretical statements (the instrumentalistic conception of scientific theory: theoretical statements express neither true nor false judgements; they do not concern any real states of affairs but, like "ordinary" analytical statements, they only constitute an instrument for "carrying out calculations" over true perceptive judgements — the judgements in the strict sense of this term).

On the other hand, such an originary and presuppositionless cognition for phenomenologists is eidetic cognition. It is presuppositionless by definition because its defining condition is a previous realization of a phenomenological reduction, i.e., the elimination of all forejudgements. However, while the presuppositionlessness of phenomenologists' eidetic cognition is much more clearly revealed than the presuppositionlessness of positivistic cognition, based directly on experience, the originarity of the former is, on the contrary, less visible than the originarity of the latter (let us have in mind that we do not mean here the originarity in the phenomenological sense). The former originarity is of a different kind — it is of a more indirect character. From a positivist's point of view, judgements directly based on experience justify all other judgements taken into consideration because of the appropriate logical relations binding the former and the latter. Phenomenological eidetic cognition, on the other hand, can, by no means, substitute for the "ordinary" (positivistic) cognition. It only evaluates critically "ordinary" cognition, and determines its (limited) cognitive significance. Generally speaking, the originarity of eidetic cognition consists in the fact that, having obtained it, we take a standpoint from which we are able to evaluate epistemologically all other judgements.

The positivistic direct experience or the phenomenological eidetic intuition constitute a foundation for the appropriate forms of naïve scientism, while Bergsonian irrational intuition or Diltheyan understanding perform the same role within the respective versions of naïve anti-scientism.

As we have already said, it is possible to distinguish at least three practically occuring forms of any anti-scientism. These are: (a) anti-scientism, which is opposed neither to the scientistic understanding of truth, nor to the fact that this truth represents the only possible type of cognition; it is opposed only to the fact that obtaining this truth is the fundamental value; (b) romantic anti-scientism, which, to conceptual cognition considered to be possible, opposes intuitional-irrational cognition as an alternative and, at the same time, more significant cognition; (c) radical anti-scientism, which opposes the claim that any type of conceptual cognition is a cognition in the strict sense. Dilthey's standpoint represents romantic anti-scientism, since, besides conceptual cognition, this philosopher also distinguishes intuitional-irrational cognition based on understanding. According to Dilthey, the latter exceeds conceptual cognition even in respect of the certainty of judgements obtained in this way (and so, also in respect of cognitive value).

Bergson's standpoint, on the other hand, represents radical anti-scientism. This French philosopher assumes that conceptual cognition is not, in fact, a cognition at all; it is merely an instrument useful for practical purposes. The true cognition can be obtained through irrational intuition only. We can thus say that naïve anti-scientism — a counterpart of naïve scientism — has three different characteristics depending on which of the three variations of anti-scientism we mean. In the case of variation (a) of naïve anti-scientism, we can speak of a non-cognitive anti-scientism because it does not formulate claims on the cognitive questions. It acknowledges a possibility of a spontaneous and indepen-dent of culture (not determined by culture) experiencing of both oneself and the world. This experiencing is either the only "ultimate" value ("the value in itself"), or an indispensable instrumental means necessary for this value to be realized. Such an experiencing is of a non-cognitive character, which means that it is neither a cognition of whatever, nor cognition is "embedded" in it. The equivalent of the cognitive pre-suppositionlessness would be here the spontaneity and the fact of being independent of culture (this concept is superior in its range to the concept of the presuppositionlessness) that are both characteristic of this ex-periencing. The equivalent of the cognitive originarity would be its, so to speak, "value-creating" character (which is an "ultimate" value or an necessary origin of such a value).

The naïve version of romantic anti-scientism, on the other hand, is oriented cognitively, just like the naïve version of radical anti-scientism. In both cases, a possibility of presuppositionless and originary, intuitive-irrational cognition is assumed, and it is supposed either to make possible a cognition parallel to (less valuable) conceptual cognition (naïve romantic anti-scientism, like that of W. Dilthey, for example), or to make any real cognition possible (naïve radical anti-scientism, like that of H. Bergson, for example). In the latter case, conceptual cognition is considered to be such a far-reaching deformation of the real intuitive-irrational cognition, that it is either denied to have any relations with the originary cognition, or it is treated as a poorly drawn caricature of the letter.

We may thus say again that naïve scientism finds its expression in the addition of statement (4) the one concerning the existence of both the presuppositionless and originary conceptual cognition — to the con-stitutive statements (1) - (3), characteristic of any form of scientism. Naïve anti-scientism consists either in completing the negation of state-ment (3); or with a thesis on the possibility of a spontaneous, non-cultural and "value-creating" experiencing of oneself and the world — non-cognitive anti-scientism; or completing the negation of statements (2) and (3) — romantic anti-scientism; or completing statements (1 — 3) — radical anti-scientism — with a thesis on the existence of a non-conceptual, presuppositionless, originary, intuitive-irrational cognition. As far as the sophisticated forms of scientism and anti-scientism are concerned, we must say that they are characterized by, respectively, rejecting statement (4), or rejecting the appropriate thesis of naïve anti-scientism.

II. THE SOPHISTICATED VERSIONS OF SCIENTISM AND ANTI-SCIENTISM

Some very significant arguments against the naïve versions of scientism and anti-scientism have been formulated by the continuators of these two orientations, who, however, have accepted them in their sophisticated forms. And as a matter of fact, the arguments developed on both "sides of the barricade" against the naïvety of the predecessors are often convergent. Hence, H.-G. Gadamer, for example (an advocate of anti-scientism), can refer to some of the theses put forward by K. R. Popper (an advocate of scientism). In order to eliminate any misunderstanding, it is necessary to add that, although various philosophers mentioned in chapter (I) usually represent the naïve forms of scientism or anti-scientism, some of them happen to be the representatives of the sophisticated forms of these orientations. This remark concerns, above all, E. Cassirer, whose standpoint is, in my opinion, closer to sophisticated scientism. The representatives of the naïve versions of both orientations are, in my opinion, those philosophers who have been explicitly discussed in the last part of chapter (I), or those, who rank among philosophical conceptions explicitly analyzed there.

1. Hypotheticism as Sophisticated Form of Scientism

One of the basic ideas of hypotheticism is the one attacking the following constitutive premiss of the naïve version of scientism: there exist judgements which are directly based on experience and which represent presumptionless and originary cognition. Orthodox hypotheticism, the representative of which is, for example, I. Lakatos (but not P. K. Feyerabend) is mainly opposed to the conception of the presuppositionlessness of the cognition which is considered by positivists to be based directly on experience. This standpoint not only allows a possibility of questioning every judgement which, according to positivists, is based on experience, but also allows to reject it only because it disagrees with the judgement which could not be in any case considered as based directly on experience (because it is a "theoretical" judgement). From the above follows that also the originarity of positivistic direct cognition must be questioned.

It must be said here that this idea is not exposed by orthodox hypotheticism which maintains a certain equivalent of the conception of originarity of empirical cognition. This equivalent is represented by the postulate which seems to be respected in empirical sciences. It reads as follows: representatives of these sciences should come to terms about observational judgements. It has not been until P. K. Feyerabend that this question was made the starting point of the opposition towards orthodox hypotheticism. A symptomatic expression of this opposition is the title of one his papers: "Science without Experience".

Within hypotheticism, the equivalents of positivistic judgements (statements) directly based on experience are so-called basic statements (judgements). The point is that these basic statements have won their place within science not because they concern states of affairs perceived *in Anschauung*: such an *Anschauung* is often experienced by an individual, which is the reason for which he accepts a given basic statement. However, these empirical *Anschauungen* can convince only that individual who has experienced them personally. On the level of a community, they constitute no argument at all. There are no reasons on the social level that would justify the acceptance of a given basic statement except for one general reason, namely that the majority of scientists agree to give up any further support of that statement. However, irrespective of whether the majority of scientists agree on accepting a given basic statement, or whether they question it, it can always be rejected by pointing out to the fact that it disagrees with certain already accepted non-basic statements, particularly those which imply that there are reasons for which one may doubt that a supposedly observed state of affairs, being a semantic referent of a basic statement, really occured.

Therefore, the fact that basic statements are of an observational character cannot guarantee that the judgements expressed by them are of exceptional certainty. The truth is that, particular scientists representing empirical sciences must (according to the rules of scientific procedure) in the first place consider their opinions within observational sphere, and in the second place — in the non-observational sphere. This second place does not mean, however, that they are less significant. In order to accept given observational statement, an acceptance of certain other, non-observational statements (non-basic ones) is logically necessary; these previously accepted (in the logical sense) non-observational statements constitute the presumptions of observational cognition. If they are questioned, then it is also possible to question such basic statements that, in a "perceptive" — empirical sense, seem obvious to particular individuals.

The arguments of hypotheticism against the positivistic conception of presuppositionless judgements directly based on experience, can be reduced to, on the one hand side, the fact that there is no never-failing relation which justifies and connects an empirical *"Anschauung"* experience with a given basic statement. The relations that could be taken into account are of an exclusively subjectively-psychological character, while the basic statement claims something "objective". It does not concern the *"Anschauung"* experience but rather the state of affairs which can occur only when certain other, presently not directly given, or even impossible as directly given states of affairs, occur. (When I say: "I can hear a thunder", for example, I do not simply register a certain acoustic experience but I ascertain a state of affairs which determines, in particular, an appropriate theory of electrical statics that explains this acoustic perception). On the other hand, the above-mentioned arguments indicate, in a more general sense, the impossibility of proving the reliability of the specified criteria of true knowledge. We were to prove their reliability referring to certain reliable metacriteria. The latter, however, are in the same situation as the former, etc. Briefly speaking, neither the positivistic directly-observational cognition is always presuppositionless, nor any kind of pressuppositionless cognition is possible at all. And the method (characteristic of hypotheticism) of indicating that basic (observational) statements are grounded on non-basic (non-observational) assumptions makes us think that neither the positivistic directly-observational cognition is always originary, nor any kind of originary cognition is possible at all.

2. The Sophisticated Forms of Anti-scientism

The anti-scientism which is not interested in the cognitive question, or, in other words, the one which does not negate the first two of the three constitutive statements of scientism but which, on the other hand, questions only statement (3), is obviously not represented by any specified philosophical orientation. Such a standpoint, however, is, beyond any doubt, met today outside philosophy. We may even say that it is close to, at least, certain aspects of the activity of modern philosophers, who, generally speaking, represent different versions of anti-scientism.

It must be said that this standpoint, liberated from a naïve belief concerning certain primitive, non-cultural, "value-creating" experiences, and thus being limited to the opinion that conceptual cognition is not a fundamental value, and that such a value is represented by a non-cognitive experiencing of certain states of affairs possible to be distinguished, seems to be quite acceptable if it is included in the framework of an appropriate relativization. Let us, however, analyze the philosophically "registered" variations of modern anti-scientism.

It is characteristic of all these variations (I mean here only these variations that are significant in modern philosophy), that their criticism is not directed towards the conceptual cognition of scientism but towards "objectivized" thinking. Objectivized thinking is characteristic not only of science (positively including the natural sciences) but also of metaphysics and the practical, common-sense view of the world. It consists not in simply using discursive concepts, but using them as consciousness (subjective) equivalents of objects, which are opposed to those objects. We may say that it consists in assuming the opposition between consciousness and the objective correlate of consciousness. Modern anti-scientism, contrary to traditional anti-scientism, does not prefer the non-conceptual, intuitive-irrational cognition; it prefers the cognition which does not respect the distance between consciousness, a concept, and their object. It is a cognition which, in a peculiar way, "synthesizes" consciousness with a conscious object. It can use concepts; the only condition that it has to follow is that it does not introduce a barrier which separates concepts from objects "thought" by it.

And so, we can say that modern philosophical anti-scientism prefers the "disobjectivizing" cognition rather than the intuitive-irrational one. It propagates this preference referring to various philosophical traditions. It speaks of ancient Greek philosophy, where no attention was drawn to a clear distinction between truth as a property of judgements and truth as the really occuring states of affairs (there is no clear difference between a judgement and the state of affairs ascertained by it); it also speaks of Hegel's *Phänomenologie des Geistes*, for which the objective reality of particular states of affairs, ascertained by their direct consciousness, is only "mediated" manifestation of consciousness represented by a critical reflexion over the former; it speaks, too, of a peculiarly Marxist concept of criticism, which refers particularly to the direct consciousness of objects of the capitalist production — treating its imaginations as an objective reality; finally, it speaks of the Husserlian diagnosis of "the crisis of European sciences". It concerns the sciences (basically the mathematical and natural sciences) which consider the natural facts that they ascertain as objective, while, in fact, they are only a result of a biassed selection of such states of affairs in which we are interested because of technical and useful reasons — without being conscious of their "sense", i.e., the way in which they are given and objectivized.

These traditions, most radically and expressively interpreted by the Heideggerian anti-Cartesianism (which opposes the dualism of object and consciousness), are, to a different degree, taken into account by various variations of modern anti-scientism. They are, however, always inspired by Heidegger's conceptions. Also symptomatic here is the fact that both Heidegger's philosophy and the ideas of modern anti-scientism which continue it are represented either by philosophers trained in naïve phenomenological scientism (for that reason we may speak of, somewhat paradoxical, postphenomenological anti-scientism), or by philosophers who at least, respect the results of Husserlian thought.

Having stressed the fact that modern anti-scientism, which does not automatically mean: sophisticated anti-scientism, wants to question not as much the conceptual cognition — considered by the advocates of scientism absolute even in the axiological sense (statement (3)) — as the objectivizing cognition, we are not, however, going to follow this articulation in the present work. The articulation that we have presented here assumes that every type of "objectivizing" thinking is of a conceptual character, but not all conceptual thinking must necessarily be objectivizing. Therefore, it is possible to think conceptually and, at the same time, in a "disobjectivizing" way. The question, however, is whether conceptual thinking (the type of thinking articulated in the form of judgements), which, at the same time, would be "disobjectivizing", is possible at all.

If we assumed that the answer to that question is positive, then modern philosopher, who, in any case, supports the "disobjectivizing" thinking and cognition, would not be — as a matter of fact — an anti-scientism advocate in the above sense. However, the very logic itself excludes the possibility of assuming the "disobjectivizing" thinking to be conceptual and, for that reason, it justifies the assumption that the opponents of the "disobjectivizing" thinking are advocates of anti-scientism. The reasons for that are the following: if the "disobjectivizing" thinking, being a conceptual type of thinking, consisted of judgements, then it would have to consist of such judgements that would ascertain both the definite states of affairs and its own way of judging them. Such judgements would then be at the same time object judgements and metajudgements. However, they are logically "forbidden": the concepts of an object language and those of a metalanguage "must not" be used at the same time within a given system of judgements, and particularly within a given act of passing a judgement, under threat of falling into a contradiction, as it is stated by modern logical semantics.

This kind of argument supporting the logical invalidity of "disobjectivizing" thinking, i.e., the one claiming that this kind of thinking cannot be a correct conceptual type of thinking, can convince only those who have decided to respect logic in its modern form, rather than the Heideggerian idea of anti-Cartesianism. Incidentally, this idea has become possible to be articulated thanks to the Husserlian conception of the eidetic "experiencing" of the sense of a given object judgment, i.e., the way in which all states of affairs, "multiradiantly" ascertained by that judgment (in different contexts of space and time) and being its intentional correlates, are "experienced". If we assume that, passing a given object judgement, we think in a synthetic way (both the given state of affairs and the sense of that judgement) — which can also be interpreted as a symultaneous ascertainment with the help of that same judgement (of a given state of affairs and the sense of that ascertainment) — then we shall be dealing with a judgement which is both object and metajudgement (it refers both to something and to itself). Being paradoxical a link between the Husserlian scientism and the Heideggerian anti-scientism, this conclusion is definitely not logically unavoidable (especially, if we assume that it is possible, in this context, to speak of some at least sketchy logical determinations). The fact that the sense of judgements can be "seen" in their intentional objective correlates does not force us to accept the specific "synthetic" character of those judgements — their objectively-metalinguistic character. We may assume that corresponding to such a "synthesis" are, despite all, two separate judgements: an object judgement and a certain specific metalinguistic judgement concerning the "epistemologically bracketed" object judgement that has been passed earlier. The so weakened conception of "disobjectivizing" thinking or cognition seems to express — first having been precisely specified — a certain valid intuition. We shall speak about it later and now, let us only add that many of the separate statements of contemporary anti-scientism advocates can be interpreted in the spirit of this weakened conception of "disobjectivizing".

And yet, it remains a fact that both Heidegger and the contemporary philosophical thought, which continues his basic ideas, declare a strong, anti-Cartesian version of the "disobjectivizing" conception. Therefore — since we have decided to distinguish between the object language and the metalanguage as a necessary condition for conceptual cognition — we shall consider this though as belonging to anti-scientism in the sense established earlier. The "objectivizing" cognition, which, according to our earlier assumptions, is the conceptual cognition, will be opposed to a somehow "better", non-cognitive, "disobjectivizing" cognition; it may even happen so that the former will be completely disqualified as a form of cognition. For that reason, we are dealing here with a new, modern version of romantic anti-scientism, or even radical anti-scientism.

The question, however, is whether these modern versions of anti-scientism assume a naïve or sophisticated form. I presume that a general answer to that question is impossible. It varies according to whether the "disobjectivizing" cognition is recognized as having a presuppositionless and originary character or not. In other words, the problem is whether the anti-Cartesian *Dasein* "brought into memory" as *In-der-Welt-sein* is identical to the "forgetting of" the world's objectivity (presuppositionlessness), which is never completely realized, or whether the fact of being brought into memory is considered as theoretically impossible without, at least partially, "forgetting the being" (the impossibility of "bringing the being into memory" without respecting certain "objectivizing" prejudgements). Is the fact of "bringing the being into memory" supposed to constitute an origin of all cognition, or of this kind of cognition which is "better" than any other type (originarity), or does the significance of "bringing the being into memory" as a process and, at the same time, a procedure, consist in something else? (Let us have in mind that both "bringing the being into memory" and "forgetting the being" are anti-Cartesian categories, which denote the peculiar ontological-epistemological wholes: processes and procedures at the same time). In any case, I think that the two probably most popular currents within contemporary anti-scientism: the one represented by the Frankfurt School (J. Habermas approach) and the one represented by Gadamerian approach are moving further and further away from the naïve form.

Within J. Habermas's epistemology, each of the types of cognition distinguished by that philosopher: the analytic-empirical one, the hermeneutic one and the critical one — assumes a certain "interest", which respectively is — the technological control, communication and emancipation. There is no cognition without such an interest, i.e., without the occurence of a certain, usually not realized, need for which particular individuals ascertain their image of reality. The illusion that it is the other way round and that a "purely theoretical" cognition which lacks the interest is possible (such an illusion is usually present in the analytic-empirical sciences like the mathematical-natural sciences and the humanities based on natural sciences; it is usually supported by positivism and the Husserlian phenomenology) performs the role of concealing the interest of the analytic-empirical cognition. From the point of view of the technological interest of controlling the nature (and possibly the society), the only useful cognition is the "objectivizing" cognition which determines the world as an arrangement of objective phenomena, regularities and processes. A natural consequence of that is that the analytic-empirical cognition is not presuppositionless.

The "disobjectivizing" cognition, i.e., the critical one or the hermeneutic-critical one, serving the emancipating or the communication-emancipating interest (in the latter, it means the communication which is "free from disturbances" — without the mental compulsion imposed upon individuals), may also not be of a presuppositionless character. However, in order for it to be free of determinants of "objectivizing" thinking, which constrain the undisturbed communication, it must be a result of "criticism", i.e., a specific genetic analysis of prejudgements, revealed with its help and peculiarly discredited. If such a "criticism" requires its own premisses, then even the most basic acts of "disobjectivizing" cognition must refer to logically earlier presuppositions.

The above interpretation of J. Habermas's epistemology determines that critical cognition does not encompass any sphere of presupposition-less knowledge. The reason for that is that it is always based on certain prejudgements, what is paradoxical since one of the features that define critical cognition is its emancipating role; it offers freedom from any prejudgements. It seems, however, that quite possible is the other interpretation — unanimous with the intentions of the Frankfurt School — according to which critical cognition does not assume any prejudge-ments at its basis. Such an interpretation seems to be justified, especially that, as a matter of fact, the criterion of validity of that cognition is represented by its efficiency as a means of emancipation. With the latter interpretation, the anti-scientism of the Frankfurt School would not be internally contradictory and would rather tend towards the naïve form.

By and large, in both cases we are dealing with a romantic type of anti-scientism because "objectivizing" (conceptual) cognition, the analytic-empirical one, is not denied the ability to cognize, but is only contradic-ted with a "better" type of cognition — the "disobjectivizing", critical or hermeneutic-critical cognition. It is not better in the cognitive respect (as it was claimed by Dilthey) but in the ethical respect: emancipation and emancipated communication (i.e., the interests that are served by that cognition) are ethically "better" than the technological control over nature and society. Similarly, the methodological anti-unism (as a normative orientation: the humanities must be excercised as hermeneutic-critical sciences) of the Frankfurt School may be said to be not of an epistemological (Dilthey) but rather purely ethical character; it is agreed that hermeneutic-critical sciences should serve the interest of emancipa-tion as well as that of emancipation and communication, thus rejecting the analytic-empirical approach. The reason is not so that they are able to cognize in a better way but that they serve "worthier" values.

Let us also notice that even if we accepted the presuppositionless character of particular fragments of critical and hermeneutic-critical cognition, they still would not be of the originary character. It is not only that criticism and critical understanding do not justify analytic-empirical statements but, what is more important, they do not even justify the predicates which are thematically connected with them. They only "persuade" them, rather. The emancipating as well as emancipating and communicational efficiency is to be one of the criteria of their validity.

The problem of originarity in the conception of "disobjectivizing" cognition connected with H.-G. Gadamer's hermeneutics is yet different. Hermeneutics assumes that all cognition, or, to be more precise, the cognizing being, comes from one origin; it is the pre-reflexive *Dasein* as *In-der-Welt-sein*. This "experiencing of being" — in the specific epistemo-logical-ontological sense — is pre-reflexive not in the sense that reflexion, or the consciousness of "experiencing" an intentionally given state of affairs, appears only after a certain intentional mental act has terminated, but rather in the fact that it is not separately concentrated on the reflexion itself, i.e., on being aware of "experiencing" and the way this experiencing goes on. It is not concentrated on the sense of what is being experienced alone (let us remember that the phenomenological sense means the way in which an intentional correlate of a mental act is given). The sense of the "experienced" state of affairs is present in mental act itself, within which this state is "experienced", and so reflexion (which grasps the sense) is already present in the pre-reflexional "experiencing". The latter is, at the same time, of an originary character in relation to all the non-originary cognition, i.e., the reflexion, which pre-selects the senses that come into being, particularly the "objectivizing" senses.

From the point of view of H.-G. Gadamer's hermeneutic philosophy, also the appropriately practised humanities are a territory on which originary cognition is obtained. The reason is that within it, one does something that he normally does in life — "simply being", i.e., performing the pre-reflexive "experiences". This "being" or "experiencing" is most extensively accumulated within language — in speach acts (in a broad understanding of the word "speaking", because one can also "speak" with the work of art, for example) and acts of understanding. In this way, the idea of the methodological anti-unism is continued: the hermeneutically practised humanities yield originary and "disobjectivizing" cognition, while the natural sciences yield an incomplete, derivated, one-sidedly "objectivizing" cognition. This is an idea similar to that of Dilthey; the only difference is the fact that the basic humanistic operation is not understanding in the sense of a mental identification with the speaking subject (who produces "the expressions of the spiritual life"), but "embedding" both speaking and pre-reflexional "experiencing", which represents particular senses.

While humanistic cognition in the understanding of the Frankfurt School cannot be originary in any of its aspects because it should be critical (this fact has been vividly demonstrated in the controversy between Habermas and Gadamer), and yet, it can be considered as presuppositionless, Gadamer's hermeneutics, nevertheless, positively excludes the latter. The pre-reflexional "experiencing" — as the territory on which "the openness of being" and "bringing the being into memory" are revealed — is always, at the same time, the territory on which "the concealment of being" takes place. There is no absolute cognition; Hegel's absolute knowledge is impossible. The pre-reflexional experience, which does not respect any prejudgements, would never be able to come into being (this idea is one of the most vivid manifestations of the fact that the postphenomenological hermeneutics moves further and further away from Husserl's approach). If, then, the representatives of the Frankfurt School believe that one can free himself by way of critical reflexion, this is only a naïve illusion. The reason is that criticism is impossible without its own prejudgements.

It can be clearly seen that both the romantic anti-scientism of the Frankfurt School and the radical (since the "objectivizing" cognition of the natural sciences is secondary and incomplete) postphenomenological anti-scientism of Gadamer move away from the naïve form. In case of the former, the reason is that the idea of originarity of all cognition (and, maybe, even the idea of the presuppositionlessness, contrary to Gadamer's opinion) is given up. In the other case, the reason is that even the most originary consciousness is not claimed to be absolute, presuppositionless and not based on any prejudgements.

III. METASCIENTISM

A common feature of the sophisticated forms of scientism and anti-scientism is the questioning of the possibility of existence a both presuppositionless and originary cognition; irrespective of the ways of comprehending this cognition within the scope of modern philosophical conceptions. We mean here such a questioning that would refer to a fairly differentiated argumentation. I think that this negative thesis shared by both philosophical orientations cannot be rejected, even though, according to my opinion:
(1) not all of the arguments used "on both sides of the barricade" and meant to support this thesis are convincing; (2) the above thesis can be formulated in a more general way viewing cognition as a specific result of a culturally regulated activity; (3) such a more general formulation is possible within the conception of culture which allows, at the same time, to establish both which "interests" (to use Habermas's term, although not quite in the way he meant it) are represented by the scientists and which are represented by the anti-scientists. The present part of my considerations will open with an introduction of an appropriate conception of culture. Then, having generalized the negative thesis that we are dealing with now, I shall present the characteristics of the "interests" of scientific cognition and the "interests" of both scientism and anti-scientism. Finally, I shall present the point of view which is contrary to both approaches.

1. Culture

Man's cultural activity is understood here as all actions which are commonly regulated within particular communities (social groups and societies). They are regulated by the fact that particular subjects of these actions respect the normative judgements, which specify the purposes of these actions (the positive values which are worth being realized), and the directive judgements, which specify the actions that lead to the realization of these purposes (in the form of either sufficient or necessary condition). Thus, man's cultural activity is more or less the same that "culture" in the ethymological sense, i.e., a systematic — commonly regulated — "cultivation". I do not, however, refer the concept of culture to that "cultivation" itself, but rather to the system of normative and directive judgements that regulate it. This is not an arbitrary terminological decision; there is a number of substantial reasons for which culture is identified with the normative-directive system of judgements, and so, with the socio-subjective regulator of "cultivation". I will not discuss these reasons here, mainly because it is impossible to analyze them precisely in a brief way without the earlier introduction of an appropriate theory of scientific cognition. Some of the elements of that theory will be presented later but only to such a degree that is necessary to present the sketch of the approach opposed to scientism and anti-scientism.

Having accepted the thesis that the above described man's cultural activity ("cultivation") is a social practice, and so (according to the definition of the latter) it is a hierarchical functional structure, determined functionally by the "requirement" of a permanent reproduction-transformation of objective conditions of a social being in the form of relations and productive forces, we approach culture as a socio-subjective regulator of the social practice. This regulator is determined and changes historically in the way adjusted to the appropriately concretized basic function of the social practice. The adjustment means that both the normative and directive judgements, commonly respected in particular communities and constituting the culture of these communities, "must" (in the sense of functional determination) be so that, respecting them, the social practice as a whole performs its basic function in its appropriate historical versions. Since all beliefs, commonly respected within the given community and, at the same time, functionally adjusted in the above way to the "tasks" of the social practice, which explains their universality, are included by me in the sphere of social consciousness, culture can be easily seen to constitute a certain (in many respects even the most significant) sphere of social consciousness.

As far as the (mentioned above) concept of functional determination is concerned (the terms "requirement" and "task" are understood here as metaphorical names of a functional determinant), I shall, at least, say that I distinguish it both from the causal determination and from the teleological determination (in all its meanings). The reproduction-transformation of objective conditions by the social practice is neither the cause for its formation and its historical transformations, nor a purpose of this formation and these transformations. We say that it constitutes a functional determinant of the social practice in the same sense in which a biologist says that a constant supply of arterial ("fresh") blood for an animal organism is a functional determinant of the functioning of heart.

There are several different types of the social practice, of which the functionally superior one is the "material" practice — the basic type of the social practice — which includes the production, distribution, and consumption of goods. This means that the basic type of the social practice is directly functionally determined by the above-mentioned determinant of the totality of social practice, while the other types of this practice participate (in a less or more indirect way — indirect functional determination) in the "task" realized by the basic practice. Both the repertoire of the types of the social practice and the scheme of their functional dependencies are, at the same time, subject to historical modifications. The only constants are: (a) the presence of the basic practice, and (b) the less or more indirect functional dependence of the other types of practice upon the "tasks" realized by the basic practice.

The set of the socio-subjective regulators of particular types of the social practice constitutes the culture of that community, or, to be more precise, the set of domains of that community culture. From the cultural perspective, particular actions which comprise a given type of the social practice are subjective-rational actions, or, in other words, actions that realize definite values, i.e., purposes in the way determined by specific directive beliefs. Irrespective of that, however, the system of normative-directive judgements is determined by the function performed within the social practice by the respective type of practice. One should add that the respect for these judgements paid by particular participants in a given type of the social practice makes their actions the subjective-rational ones.

The formation and the transformations of a given cultural domain depend not only upon its effectiveness as a socio-subjective regulator of the social practice, functioning in a specified way, but also upon its hitherto condition, or, to use Engels's terminology, its hitherto "thought material" represented by that discipline. Assuming a form new in relation to the hitherto one, each cultural domain establishes, at the same time, its relation to the past form. The new form is determined not only by the new functional "requirements" aimed at the type of practice regulated by that domain, but it is also limited by the possibility of establishing its "thought" relation to the previous form (continuation, correction, negation). Having that in mind, we may say that subsequent developmental stages of particular cultural domains are determined in the functional-genetic way.

All cultural domains are grouped within two basic spheres of culture: the technological-useful one and the symbolic one. The sphere of technological-useful culture is constituted by the normative-directive system which regulates the basic type of practice. A characteristic feature of the cultural domain that constitutes this sphere is that the appearing of the effects of actions which constitute the practice regulated by it and which belong to the range of functioning of that practice is not necessarily dependent upon the common respect for the normative-directive elements of that domain as beliefs characteristic of the subjects of these actions. In the case of the large group of symbolic culture domains, it is on the contrary: a necessary condition for the appearing of the effects which belong to the range of functioning of practice types regulated by these domains, is the common respect for their elements as beliefs characteristic of the subjects of actions comprising particular practice types.

The above distinction seems a little intricate. However, I am sure that a simple example will indicate that it is, in fact, not so complicated. Let us take the example of the technological-useful practice, within which takes place the activity that consists in using a T.V. set (produced within the same practice). To what extent will the T.V. set holders recognize the norms and directives which regulate the course of its production as beliefs determining the actions of the producers, has no influence upon the technical qualities of the picture on the screen. And so, the T.V. set functions (technically) irrespective of whether its holder respects the normative-directive beliefs which he ascribes to its producer. The situation is completely reversed in the case of — let us quote the possibly simplest example — a sentence uttered by anyone within a given language. The effect of the action of uttering this sentence — the action which comprises the practice of a linguistic communication — belongs to the range of functioning of that practice and, as such, is dependent upon whether the receiver respects the linguistic (grammatical and semantic) norms and directives, considered by him as norms and directives characteristic of the subject of the utterance. If the receiver is not able to respect them, for example, when he does not know that language, the communicational effect (belonging to the range of functioning of the practice of linguistic communication) will not occur at all.

I think that the above-mentioned difference between the sphere of technical-useful culture and the sphere of symbolic culture, which, at the same time, allows one to contrast the symbolic-cultural practices with the basic practice, makes it possible to properly understand Marxist intuition in this respect. This intuition is manifested by the application of the term "material" to the basic practice (or, to be more precise, in applying the term "historical materialism": what is material does not require, in order to come into being, to pass through consciousness; it is not the simple physicality as it is often understood).

There are two groups of symbolic culture domains. The first group is distinguished by the fact that actions composing the practices regulated by the cultural domains belonging to it, are oriented towards the practically attainable values (as purposes). Practical attainment of values meant here relates these cultural domains with the domain of technical-useful culture. I would list here, first of all: language, art, custom, science, and social political-legal consciousness. The other group of symbolic culture domains is characterized by the fact that the values, the realization of which they subjectively regulate, are of an "ultimate" character. It means that these values are never presented as the means of realization of any other values and are of a worldview character (we shall return to that problem later); most often, they are practically attainable. The set of symbolic culture domains included in the first group, can be called symbolic culture in a narrow sense, while the set of symbolic culture domains included in the second group, can be called the worldview sphere of (symbolic) culture. The types of the social practice subjectively regulated by these domains may be called worldview-creating practices. These include: the practice of magic (magic as a cultural domain), religious practice (religion), secular worldview-creating practice (secular worldview).

2. Assumptions of the Semantics of Cultural Communication as "Cultural A priori"

Science, as it has already been stated, is understood here as one of the domains of culture. It is a socio-subjective regulator of a particular type of the social practice — the scientific practice. It can also be understood as a system of normative-directive judgements, constituting the social methodological consciousness of the scientific practice. The social methodological consciousness is appropriately differentiated in the context of particular branches of that practice and its different stages of historical development. It is obvious that if we assume the scientific cognition to be a set of judgements "produced" within the scientific practice in a culturally regulated way, then the reflexion on this cognition must begin with accepting certain general theses that belong to the theory of culture. The so understood theory of scientific cognition might seem to be simply a certain part of the theory of culture. This problem is much more complicated, however. It is a fact that the starting point of our theory of scientific cognition is the theory of culture; let us then call the theory of scientific cognition which assumes the above conception of culture *historical epistemology*. The point is, however, that, thanks to its studies on the history of science, and particularly on the history of the scientific reflexion on culture, historical epistemology provides for its mother discipline (the theory of culture) a number of methodological directives making it possible to significantly broaden its initial state of cognitive possession. Neither these directives nor the cognitive effects obtained with their help are consequences of the general statements of the theory of culture, that were assumed earlier. And so, the relation between our theory of culture and historical epistemology can, by no means, come to the following scheme: a general system — a particular case. It rather resembles a relation which, in cybernetics, is called a feedback.

Certain premises and establishings specific for historical epistemology will be discussed later. Before that, we shall try to generalize in terms of the above outlined conception of culture a common thesis of the sophisticated versions of scientism and anti-scientism, i.e., the one saying that there is no originary and, at the same time, presuppositionless cognition. Then, having discussed the selected claims of historical epistemology, we shall define, on the basis of these claims particular "interests" of scientism and anti-scientism.

Particular domains of symbolic culture in the narrow sense consider the acts of communication of particular states of affairs either as superior goals of actions creating social practices (e.g., language, tradition, art) regulated by these domains, or as the necessary means of realization of the respective superior goals (e.g., science or political-legal consciousness). This communication — be it linguistic, artistic, scientific, or based on the customs — is regulated by the semantic rules (or, to be more precise, semantic directive judgements) specific for each of these domains. These semantic rules assign to particular actions or to their products, i.e., communiques (such as linguistic, artistic, scientific, based on the customs etc.) — specific referents. Thus, each domain of symbolic culture in the narrow sense includes a certain set of semantic rules. We shall call it semantics. And so, we shall obtain, respectively: the semantics of linguistic, artistic, scientific, etc. communication.

Logically as well as historically prior to each semantics of this type, are certain judgements which determine *a priori* and characterize the states of affairs assigned by this type of semantics as referents to appropriate communiques. We shall call them assumptions of the symbolic-cultural communication semantics (to give an example: the semantic rule which specifies the denotation of the word "table" is determined by such assumptions according to which every x that belongs to that denotation is a physical object, consists of a "top" and an appropriate "support", and can be used in such and such a way, etc.).

Informing one another of certain states of affairs, we communicate with one another to such an extent and to such a degree of effectiveness to which we respect the semantics of communication together with its assumptions or determinants. If we did not fully respect it, such symbolic-cultural social practices as the linguistic, based on the customs, artistic or scientific practices would not be possible, and our individual actions in this respect would be inefficient. This does not mean, however, that this situation is characterized by a uniform homogeneity. As it was already mentioned before, we acquire the semantics of communication together with its assumptions, or — to be more general — culture, or — to be even more general — social consciousness — in various individual ways. We also use and respect it in such an individually varied way. This seems to be the source of the individualistic illusion of a personal-subjective perspective in which the source of culture, especially cognition, is the individual experience, while in fact, it is only a mental device with the help of which an individual is able to participate in culture or, to go even further, in social practice.

And so, at this point, the common negative thesis of the sophisticated varieties of scientism and anti-scientism can be corrected and reinforced to the following form:

The referents of symbolic-cultural communiques (a) are always relativized to the assumptions of their semantics, which (b) are subject only to a partial correction and revision.

Part (a) of the above thesis indicates that no socially communicable, ordinary or scientific, cognition can be free of assumptions, i.e., presuppositionless. Scientifically ascertaining certain states of affairs, we determine a large number of all the assumptions of the semantics of that ascertainment and, at the same time, we are not able to be conscious of all these assumptions. We can only point our attention to a certain fragment of them. The reason for that is not the large number of these assumptions but the fact that the ascertainment of determining a given assumption determines itself certain new assumptions with its own semantics.

Moreover, our thesis concludes that no scientific cognition (and not only it) can be originary because the individual experience — even the one perceived as most originary given — which has the respective assumptions of the semantics embedded in it, has nothing to do with it. Having rejected the individualistic perspective, we cannot accept any (individual) experience — even the "being in the world", considered as not presupposition free — to be an absolute origin of cognition. It does not change the situation even if we state that the revision of the assumptions of semantics can also be experienced or "embedded" in experience. The reason for this is that both experiencing the world according to the hitherto assumptions of semantics and experiencing its change according to the way of revising some of its assumptions are always secondary to the process taking place in culture, and particularly in science. It is secondary not in the sense of being delayed in time or that this process may occur without it, but because the respective assumptions and revisions are the characteristic features of socio-cultural states and processes, and it is in terms of these features, that we can define and question this experience, not the other way round.

3. The "Interests" of Scientism and Anti-scientism

I am using here Habermas's term "interest" (but put in quotation-marks) in order to be able to refer to certain intuitions which, because of the growing popularity of that philosopher in Poland, are more and more commonly associated with that term. As we have said, involved here are only some of these intuitions, and for that reason this term is put here in quotation-marks. According to Habermas, an interest is a need of particular individuals, who are usually not aware of it, or, in other words, it is a functional determinant of their way of thinking about the world. This way is inasmuch an "ideology", or a "false consciousness", as it does not recognize its own determinants. Speaking of an "interest" (in quotation-marks), I mean a functional social determinant of the type of practice, i.e., the "task" realized by that type of practice in a historically varied way, and specified irrespective of how the normative-directive beliefs determining socio-subjectively the actions belonging to this type of practice are manifested. Briefly speaking, Habermas's "interest" is used by me as a name of the usually unassociated functional determinants of a given type of social practice, leaving out these constituents of the original conotation of the term which are connected with the individualistic way of thinking of that German philosopher.

Considering it necessary for the final part of the present work, I shall now present some premises or conclusions resulting from historical epistemology which views science as a cultural domain, and the scientific practice — regulated by it in a socio-subjective procedure — as a definite type of social practice. The most important thing here is to specify the "interest" (functional social determinants) of that practice. Then, we shall try to answer the question which "interest" is characteristic of the scientism, and which — of the anti-scientism orientation.

It seems to be necessary here to distinguish two types of judgements which express our representations of the world. These are — on the one hand — practical-cognitive judgements, i.e., (a) directive judgements which instruct about what practically attainable activities undertaken in practically attainable circumstances lead (are sufficient or necessary) to the occurrence of respective, similarly attainable, effects; (b) judgements which are the deductive reasons for the (a)-type judgements. On the other hand, they are judgements which do not have any implications of the (a)-type, but which imply normative judgements. The second type judgements or the entire groups of such judgements will be called worldview messages.

The function of the practical-cognitive judgements is quite obvious: they constitute premisses of directives for the appropriate types of social practice. Much more complicated is the set of functions performed by worldview messages. The name applied to them is justified by the fact that they constitute specific worldview systems, i.e., the sets of judgements which (a) distinguish certain groups of "ultimate" values, or, in other words, non-instrumental superior values (positive and also negative); we shall call them worldview values; (b) define the relationship between practically attainable values and worldview values, which can be — although, most often, they are not — practically attainable. This relationship can be consistency, inconsistency, etc. One of the elementary functions of a worldview and the messages constituting it is — in relation to social practice — the valorization of practically attainable values, or, in other words, taking part in the formation of socially accepted representations connecting these values with the worldview values.

Returning, for a while, to the scientific practice, let us observe that it is determined by its practical-cognitive function of the "producer" of practical-cognitive statements as well as by the worldview valorization function of the "producer" of certain elements of a worldview. The latter function, however, is not specific for the scientific practice, as it is also performed by the worldview-creating types of the social practice: the magic, the religious, and the secular — "metaphysical" one. And so, the function that defines the scientific practice is its practical-cognitive function. The function of worldview valorization is performed by it only to such an extent to which the practical-cognitive establishings may be included in a worldview.

Having assumed such a point of view, we may observe the following: (1) The scientific practice never possesses any opportunities to create full worldview systems, as it is done by the worldview-creating types of the social practice. The reason for this is that these systems cannot originate exclusively from practical-cognitive establishings, but are forced to include normative worldview messages that are, by definition, not produced by the scientific practice. (2) The so understood scientific practice and also science as a cultural domain are terms with a much narrow meaning than its commonly used and equally sounding counterparts. The institutional practice which produces practical-cognitive establishings (originally, from the range of the mathematically oriented natural sciences) emerged during the 17th and 18th century. Therefore, we can only speak of certain genetic beginnings of science and the scientific practice in ancient times or in the middle ages. Moreover, what is called humanities, does not represent, to a large degree, the scientific practice and science in the strict, accepted here meaning of these terms. Thus, we are dealing with the scientific practice only to such an extent to which the secular worldview-creating practice of producing secular worldview messages is involved.

For the above reasons, the humanities function within the worldview valorization in a particularly exposed way. This function is dominating, particularly at certain stages of development, in some humanistic disciplines in comparison to the practical-cognitive function. This domination is sometimes so overwhelming that it almost entirely eliminates the practical-cognitive function. Besides the definitional — practical-cognitive and non-autonomous function of worldview valorization, we can also distinguish a third function of the scientific practice. It is the direct cultural-educational function. It consists in a persuasive introduction of particular individuals into culture and persuasively making them respect appropriate normative-directive cultural systems. It is partially a specific type of the valorizational function. It constitutes such a type to such a degree to which it consists in persuading specific worldviews and worldview messages.

It can be easily observed that the function of worldview valorization, connected with providing for worldview elements, is performed by the natural sciences in a much less exposed way than by the humanities. Also the cultural-educational function is characteristic of the humanities rather than the natural sciences. In this work, it is called the direct cultural-educational function because it is also possible to distinguish the indirect function of this type. The latter is performed mainly by the humanistic scientific practice in such a way that it provides for practicalcognitive establishings that can be used as a source of directive premisses for pedagogical actions, which introduce one to the participation in culture. Let me stress this again: with the help of these establishings, the activity introducing into culture is organized rather than the participation in culture is directly persuaded. Therefore, the indirect culturaleducational function is a particular variation of the practical-cognitive function.

While the functions or the "interests" of the scientific practice consist mainly in providing for practical-cognitive establishings and — somewhat secondarily — providing for worldview elements and something that we might call "persuading culture", the "interests" of the scientism and anti-scientism are defined by the fact that the following worldview messages are produced within them: (1) the ones valorizing or (and) persuading the values realized within the practical-cognitive function of the scientific practice (scientism), and (2) the ones valorizing or (and) persuading the values realized in relation with the secondary functions of the scientific practice (anti-scientism). Naturally, the "interest" of anti-scientism is broader; it is connected with the "interests" of all those types of the social practice which function in a valorizational or (and) directly cultural-educational way. However, we are dealing here only with the "interest" of anti-scientism in relation to the scientific practice and science. By and large, it is a destructive "interest" because when the secondary functions of the scientific practice begin to expand and dominate over its practical-cognitive function — this being the content of the "interest" of anti-scientism in relation to this practice — both the latter and science in the sense accepted here become disintegrated.

There is no possibility here to present more arguments showing how particular phenomena occuring in the historical development of the scientific practice and science, as well as their branches, can be explained in terms of appropriate arrangements of the three "interests" discussed above. Some of the explanatory relationships are self-evident; it is easy, for example, to understand why "communicativeness" is not required of natural scientists, while it (and even more, in the form of "a nice language") is required of the representatives of the humanities. If not humanities could not function either as a source of worldview messages making valorization possible, or as a device of a cultural-(worldview)-educational persuasion. We have also no place here to present all the relations between the above analyzed modes of the realization of the "interests" of scientism and anti-scientism and these "interests" themselves. Therefore, we shall limit ourselves to: (1) stating our attitude towards the constitutive views of both orientations from the point of view of historical epistemology, and (2) presenting a third standpoint, which we shall call metascientism.

4. Metascientism

Let us observe first that, identifying the "interests" of the two orientations discussed in the above way, historical epistemology yields no basis for estimating them. The fact that the "interest" of scientism is valorization and also persuading the practical-cognitive "interest" of the scientific practice (we shall see later that this formula will have to be specified more precisely), while the "interest" of anti-scientism is a specific destruction of this scientific "interest" in favor of creating a worldview and the cultural-educational persuasion, does not determine anything axiologically. The reason for this is that historical epistemology tries not to force its way into the range of the worldview message, which is the sphere of axiology. It does not solve the dispute between scientism and anti-scientism. It does not claim that the "production" of practical-cognitive establishings is more significant than the worldview valoriza-tion or (and) cultural-educational persuasion (because, for example, only these findings represent a useful "positive knowledge"), neither does it claim the opposite (because, for example, culture can exist without science — which is demonstrated by the history of culture — but it cannot exist without a worldview). Everything that we can expect from historical epistemology, in respect of what is of interest to us, can be expressed in the following proposals: (a) a certain revision or modification of the trends of thoughts which are of a practical-cognitive character, and (b) a more precise specification of the "interests" met by the entire set of respective worldview systems.

We have already carried out a certain explication of the common thesis of the sophisticated varieties of scientism and anti-scientism — the one concerning the non-existence of a cognition, which would be both originary and presuppositionless at the same time. In such a general and modified form, we have accepted it as an element of the theory of culture which constitutes the starting point for historical epistemology.

We shall now analyze the constitutive idea of all kinds of scientism (statement (1) and (2)) from the following formula of ours: (conceptual) cognition is true only when it is represented by the judgements that are adequate in relation with the states of affairs concerned in these judgements. We shall assume that this idea refers to the practical-cognitive judgements (establishings) of the scientific practice, or to the everyday experience. The latter will be specified as a concept later, especially that it will be necessary for the consideration of the degree of validity of our idea. It could also be defined as the idea of formal realism. The term "formal" means that it can be represented by two versions — the substantially realistic one (the one of Aristotle) and the idealistic one (the state of affairs described by a judgement does not exist independent of that judgement).

Let us notice that the thesis of realism in both of the above versions is inefficient; determining the truthfulness of judgements according to that thing assumes possessing some kind of an absolute, originary, and presuppositionless knowledge on either the "external" states of affairs that are described by all possible judgements, or on the transcendental norms, which when satisfied by the judgement guarantee its truthfulness and, at the same time, guarantee that the state of affairs described by it is real. We have no cognitive access to the so conceived states of affairs; only the semantic referents of particular judgements can, in fact, be taken into account. And so, if one determines the truthfulness or cognitive validity of particular judgements in the spirit of the inefficient idea of realism, then all he does is an unjustified absolutization of the semantic referents of these judgements, granting these referents the status of absolute reality (i.e., the "externally" or idealistically understood reality). Therefore, it can be easily noticed that certain intuitions of anti-

scientism, connected with its criticism of the scientism's "objectivization", are not invalid.

The above-mentioned "objectivization", effected in the spirit of the
inefficient idea of realism, cannot function by the practical-cognitive
procedure. And yet, (1) it somehow supports the practical-cognitive
"interest" of the scientific practice in a worldview-persuasive way, and (2)
it even contains in itself a certain component without which science
would not be possible.

Let us first analyze (1). Carrying out the worldview valorization of the
absolute reality of referents of a given cognition, and, at the same time,
persuading the truthfulness of that cognition, an advocate of scientism
becomes its defender and propagator. Whether he is, at the same time, a
defender and propagator of the practical-cognitive "interest" of the
scientific practice, can be discovered after taking into consideration
several other findings of historical epistemology, concerning the
development of that practice.

A genetic starting point for the scientific practice is the social practical
wisdom, or a set of judgements of a practical-cognitive character. These
judgements (a) are generated spontaneously (and not — as in the case of
the scientific knowledge — institutionally) and commonly respected
within particular types of the social practice; (b) imply directive
judgements directly connecting the practically attainable actions and
their circumstances with the similarly attainable effects. The scientific
practice has emerged when the practical wisdom started to gradually
lose its functionality in relation to the new "interests" (needs) of the
basic, social practice, which has developed rapidly since the beginnings of
capitalism.

The reason for a growing afunctionality of the practical wisdom in
relation to the new needs of the social practice was the unsystematized
and unprecise character of its elements and concepts intervening in it.
Therefore, the first stage in the development of the scientific practice
consists in accumulating, systematizing and precisely stating the
knowledge which stays within the cognitive horizons of the practical
wisdom. This stage is often referred to as the pretheoretical stage.

The theoretical stage in the development of the scientific practice is
dependent on and, at the same time, distinguished from the pretheoretical
stage by the fact that the knowledge accumulated within the former
remains in a significantly correcting correspondence relation with the
knowledge accumulated within the first stage, which is the knowledge
represented by the social practical wisdom. This means that (a)
theoretical knowledge views the fact of the social acceptance of
pretheoretical knowledge as a phenomenon which must be explained; (b)
this explication is effected by means of pointing to the range of effective

practical usability — pretheoretical knowledge; this range constitutes a certain subdomain of the domain acknowledged by theoretical knowledge as reality; (c) what is acknowledged as reality by pretheoretical knowledge (social practical wisdom) is in no logical relation with the domain acknowledged as reality by theoretical knowledge; what occurs here is a logical incompatibility, and so, theoretical knowledge never contradicts pretheoretical knowledge but epistemologically brackets it (as a phenomenologist would say) and undertakes only the problem of the range of its effective practical usability, which is the problem of the reasons of its social acceptance. Let us also add that the significantly correcting correspondence relation combines and, at the same time, separates the physics of Aristotle and that of Galileo and Newton, the latter and the relativistic physics, the bourgeois economy and Marxian economy, positivistic-structuralist linguistics and the theory of generative-transformational grammar, etc. These examples clearly indicate that the significantly correcting correspondence occurs as a connection not only between the pretheoretical knowledge and the theoretical one but also between particular subsequent stages of the development of the theoretical scientific practice.

Besides the significantly rectifying correspondence, we can also distinguish a "normal", "generalizing" correspondence, which is not connected with the cognitive turn. It combines two subsequent stages of the cumulative development of the scientific knowledge; the subsequent stage is a generalization of the previous stage and is usually connected with its rectifying restriction. Needless to say, both stages of the development are entirely logically compatible.

Let us now come back to the question of to what degree the inefficient scientistic realism is an "ally" of the practical-cognitive "interest" of the scientific practice. Although the answer to that question was positive, it has also been rather general. At this point, we are able to specify that answer more precisely. We have observed the practical-cognitive "interest" of science to be of a developmental-historical character; to be an apologist and an advocate for this "interest" means to be an apologist and an advocate for a more developmentally advanced scientific knowledge to the disadvantage of the less advanced knowledge. Briefly speaking, the inefficient scientistic realism serves the practical-cognitive "interest" of the scientific practice when it is progressive, when it valorizes according to the worldview, and when it persuades as an objective reality the domain which semantically refers to the "ascending" scientific cognition. Otherwise, the scientism is antifunctional in relation to that "interest". And so, for example, the application of the inefficient scientistic realism to the Ptolemaeus's conception of the planetary system — with symultaneously viewing Copernicus's conception as exclusively an instrument of a mathematically convenient astronomical calculation — is, in times of Copernicus and later, antifunctional in relation to the practical-cognitive "interest" of the scientific practice. On the other hand, however, persistently supporting the scientistic-realistic Newtonian interpretation of Copernicus's theory has become, since the times of relativistic physics, also antifunctional in relation to that "interest". In other words, the functionality or antifunctionality of the scientistic-realistic valorization and propaganda of a given system of scientific knowledge are historically relative.

It has already been said that the inefficient scientistic realism contains a certain element which must always occur also in science, in the social methodological consciousness of the scientific practice; without that element the scientific practice would not be possible. The point is that a scientist representing a given system of scientific knowledge must for logical reasons, treat the semantic referents of the statements comprising that system — in an objective way. In particular, he could not "disobjectivize" them — as it is postulated by modern anti-scientism — declare certain statements and symultaneously take them in epistemological brackets, or reflect on them in a critical way. Such actions would destroy logic, without which a conceptual scientific cognition is impossible.

Surely, a scientist does not have to be an advocate of scientism at the same time; he does not have to valorize and propagate the semantic referents of his statements as realities in a inefficient sense. Such a valorization and propaganda are not a necessary condition for an objective approach to the semantic referents of one's own knowledge, although they are a sufficient condition. On the other hand, a different and much weaker — though still sufficient — condition of such an approach would be to respect the idea that could be expressed as the following norm of progressive semantic realism: treat objectively semantic referents of the progressive system of the scientific knowledge and, take in epistemological brackets systems which are earlier in their development and are logically incompatible with the progressive system of knowledge.

Let us observe that progressive semantic realism, which already is a certain worldview message going beyond the frames of historical epistemology, has a certain advantage over the inefficient scientistic realism, this being the fact that it is efficient, i.e., operative. It means that in order to realize the norm that expresses it, it is enough to follow the correspondence directive offered by historical epistemology. Moreover, it is immuned against the arguments of anti-scientism which concern the "objectivizing" thinking and which we have acknowledged as intuitively valid.

As far as the second problem is concerned, it seems to be enough to say that progressive semantic realism is far from any form of absolutism. Neither does it finally and ultimately "objectivize" certain states of affairs or certain domains semantically referring to the appropriate systems of knowledge. Facing a new, more advanced system of knowledge, it is ready to accept the "disobjectivization" of an earlier representation of a given domain. The only thing it cannot accept is stating something and, at the same time, "disobjectivizing" it. This way of operating is characteristic of the scientific practice, taken in the context of its historical development. Therefore, the objection raised by anti-scientism, saying that natural, analitical-empirical, etc. sciences always objectivize absolutely, is a misunderstanding which results from considering scientism as the real and only possible consciousness of the scientific (natura, analitical-empirical, etc.) practice, as well as from the individualistic acceptance of the perspective of a single scientist, who represents the current system of the scientific knowledge.

Needless to say, progressive semantic realism, similar to the inefficient scientistic realism, expresses the practical-cognitive "interest" of the scientific practice. And yet, its axiological poverty (it consists of a single norm) causes that scientistic realism has an overwhelming advantage over it in respect of the persuasive-valorizational functionality. In order to even slightly reduce that distance, it would be necessary to expand the persuasive-worldview argumentation, which calls for respecting the norm of progressive semantic realism. I shall not do that here; I shall only point out to a possible direction of that argumentation, which would stay in accordance with my own worldview. And so, from the point of view of historical epistemology, it is beyond any doubt that respecting the norm of progressive semantic realism inevitably leads to respecting the correspondence directive, which promotes the practical-cognitive development of the scientific practice. Let us now assume the following norm: explicitly drawing out the axiological sense, which, to a different degree, is "embedded" in the term "practical-cognitive development of the scientific knowledge" — with the help of the term "cognitive development" (which, according to what has been postulated, is descriptively synonymous) — we may say that within the scientific practice, it is necessary to aim at the cognitive development. For that reason, progressive semantic realism is a means of obtaining such a development. The basic argument supporting the axiological worldview sense of the concept of the cognitive development is the practical-cognitive functionality of the scientific knowledge, which increases together with that development.

The so outlined fragment of worldview, which neither sets up any absolutistic-excluding pretensions (i.e. it does not support statement (3), characteristic of all scientism), nor even persists that the term "cognition" referrs only to the practical-cognitive findings (even though it stresses its different meaning, especially when it is referred to worldview messages), and so it does not maintain statements (2) and (1) (progressive semantic realism and inefficient scientistic realism are two different views), should rather be considered as a continuation of certain appropriately corrected ideas of scientism. What is most important, it could perfectly fit the purpose of expressing the same "interest" that is expressed by the progressive uses of scientism. For that reason, I have decided to call that fragment of worldview *metascientism*, the prefix *meta-* indicating the fact of its being a result of the criticism of scientism (and also anti-scientism), in the classical, Marxian meaning of the word "criticism".

Translated by Krzysztof Sawala

Leszek Nowak/Poznań

SCIENCE, THAT IS, DOMINATION THROUGH TRUTH

1. Between truth and the social service

The common-sense conception of science identifies it with the cognitive mechanism society brought into being in order to attain truth about the world we live in. There are, then, two categories science is usually referred to, viz. those of truthfulness and social service. And, correspondingly, there are two basic types of philosophical conceptions of science. According to the epistemological ones, the structure of science and the principles of its development are defined in terms of the relationship between cognitive results and reality. It is maintained that the main characteristics of scientific activity is that science yields a cognition which is certain, probable, falsifiable, possessing ever greater truthlikeness, etc. Different variants of the epistemological relation between the cognitive subject and the actual object are at work in particular proposals having, however, one thing in common: all of them refer the existing scientific practice and the results of it to the reality. If the epistemological conceptions of science mention the relation between science and its social environment at all, the only type of problems they turn out to be able to pose is what the social conditions of proper (that is, defined within a given conception) functioning of science are. In case of the functionalist conceptions of science it is on the contrary. Here the initial idea is that of the global social system and the room it offers for a particular type of social activity collecting informations necessary for the system as a whole. It is maintained that the main characteristics of science is that it serves the development of the productive forces, satisfies the specific cognitive needs of the human species, realizes definite values present in the historical heritage of the Western culture, etc; different factors of the functional nature were, and are, proposed. If the conceptions in question discuss epistemological matters at all, then they attempt to transform them into functionalist ones, e.g. defining the notion of truth in terms of social praxis.

The basic difference between the epistemological and functionalist grasps of science consists in what is considered to be the principal feature of scientific activity; either that it gives us truth about reality or that it serves the society by satisfying definite social needs. As a consequence, the science of science is placed quite differently within both the grasps. For an "epistemologist" the theory of science presupposes the theory of knowledge (or some of its specialized fields like logic), whereas for a "functionalist" the theory of science presupposes the general theory of society being a specialized branch of the sociology of knowledge. Service to truth or service to society — that is the alternative that common conceptions of science offer today.

2. Science serves itself

What seems to be, however, a little bit suspicious is that in both cases science comes off so well. In fact, it is hard to imagine a more complimentary self-assessment of the people who make science than actually this: "We are servants of the Truth", or "We are servants of the Society". One may then conjecture that both the grasps of science possess a rather ideological sense for the scientific community.

It is a characteristic thing that both epistemic and functional conceptions of science neglect the internal structure of the scientific community, treating it as a homogeneous whole. If it is discussed at all, the only reason for that is to consider the optimal (resp. the actual) division of labour between different branches or roles necessary for attaining the best cognitive (resp. practical) results that science was called into being to realize.

Meanwhile, the most elementary experience every of us acquires while working in any domain of science is that the scientific community is full of incessible conflicts, and its structure has nothing in common with the community of equal people discussing critically all the proposals on the market of ideas in order to choose the one which is rational (probable, truthlike etc) to the highest degree. In reality, the scientific community is hierarchised as much as any other — except, perhaps, some churches. In fact, it is based on inequality of people making science — a man with a "famous name" may repeat the same ideas for the tenth time and he or she will find a publisher without difficulties, while a young man may discover much deeper ideas and he or she will be ignored. The truth is that followers of different theoretical approaches fight each other not with arguments as much as through monopolization of crucial positions within the world of scientific institutions.

Obviously, every man of science knows that. But the existing ideology of scientific community tells him or her to treat all the facts as "deviations" from the ideal standard of Truth or Social Service that science "in principle" realises. In this way he or she becomes unable to recognize what actually takes place in the scientific community. Moreover, this inability turns out to be necessary condition for a successful scientific career. In the same way a necessary condition for a successful career in big business is one's inability to recognize the class nature of the economy, that is, the inequality among people as to their economic power.

For in science there always exists a priviledged minority. The initial point for a theoretical analysis of science as a social phenomenon is to be, then, a recognition of the simple fact that science — apart from the economy, the political power, and religion — generates social inequality of a kind and the basic mechanisms of its internal functioning lead to the deepening of the inequality. First of all, science serves itself.

3. Science vs. knowledge. The Idealizational Conception of Knowledge

What has been said does not mean, however, that the discussed ideas of science are senseless. Certainly, in our times it is science, first and foremost, that attains truth; the problem is, however, who does that and who possesses the interest in delaying further progress. Certainly, it is science which affords practically useful knowledge; the point is, however, who does that and who makes use of it. It is impossible to answer these question without taking into account the internal structure of a scientific community. In order to understand the social function of knowledge produced by a scientific community, the latter cannot be considered as a homogenous unit.

Obviously, this could not be said, if purely epistemological problems were posed: what is the nature of truth?, which is the optimal form of theory for attaining truth?, which is the optimal method of testing whether a theory is true? etc. Yet, by doing this one builds a theory of knowledge, not a theory of science. The theory of knowledge assumes an ontology alone and considers what knowledge is to be like if it is to reflect reality equipped with such and such ontological properties. The theory of knowledge abstracts, then, from our human — biological, psychological, social etc — cognitive limitations and reinforcements. It simply ignores the existence of the human subject in the way that geometry ignores the existence of the measuring persons.

The theory of science, instead, is a theory of human cognition made by specialized people forming professional scientific communities. One of its main tasks is to compare the actual, human cognition with the ideal patterns given by the theory of knowledge and to discover factors responsible for deviations from the model based on assumed ontology alone. Due to this, the theory of science cannot abstract from the internal structure of a scientific community.

As far as the present paper is concerned, it is the idealizational theory of knowledge (Nowak (1971), (1980)) which is assumed here. Its task may be, then, formulated thus: to analyse the internal structure of the scientific community. In this way both a contribution to the theory of science (for it is the structure of the scientific community which is the source of significant deviations from the patterns proposed by the idealizational grasp of knowledge) and to non-Marxian historical materialism (Nowak (1979), (1983)) (for the scientific community is a specialized branch of the producers of culture) is attempted to be made.

4. A class system

Having the distinction between knowledge (the perfect cognition) and science (the social cognition) made, let us come back to our initial point. According to it, science is a mechanism generating social inequality between participants of the cognitive process. It seems to be natural, then, to apply the notion of class to the analysis of the scientific community.

In non-Marxian historical materialism the notion of class is introduced in the following manner. A class system is identified as:

$$(U, \ T, \ d, \ C0, \ C1, \ a)$$

where U is a set of people, T — a set of technical means used by people of U in their activity, d — relation of control over means of T, $C0$ — the set of disposers of some means of T, $C1$ — the remaining people of U, a — alienation of people of set $C1$. That somebody controls means t of set T signifies that he or she makes decisions concerning the goals of the use of t. The class $C0$ of disposers of means of T is composed of all the persons that control some means of technology and only of such persons. The relation of controlling the means should be, then, distinguished from that of making use of them, as the technological tools can be, and even usually are, employed by members of class $C1$ executing orders of members of $C0$. Alienation of members of class $C1$ consists in the fact that their own objectives cannot be achieved in the activity they must undertake as being subordinated to some of the disposers of material means and, quite on the contrary, they must realize the goals of the disposers. This justifies calling class $C0$ a class of oppressors and $C1$ a class of oppressed.

It is the economy which is such a class system. U is a set of people undertaking the economic activity, T — productive technology, d — the relation of ownership (in the factual, not necessarily juridical, meaning), $C0$ — the class of owners, $C1$ — the class of direct producers, a — the economic alienation resulting from the exploitation of direct producers. It is also the rule which is a class system. In that case U can be identified as a set of people leading public activity, T — the technology of coercion, d — the relation of controlling of T, $C0$ the class of rulers, $C1$ — the class of citizens, a — the civic alienation (Nowak (1983)).

The culture of a given society can be conceived of as a class system as well. As a material base of the economy are the means of production and of the rule are those of coercion, means of production of consciousness are the material base of culture and, as only a minority of people leading the economic or the public activity control the material means the activities in question depend on, so a minority of participants of culture control the material means of spiritual production, making decisions of which cultural contents are, and which are not, spread with the aid of those means.

The question I would like to consider now is the following: is science a class system of a kind?

5. The class nature of scientific community

In order to find an answer it will be suitable to consider the position of somebody who joins a given scientific community.

The first experience of a young scientist is the recognition of the fact that within the community of scientists there are people who have theoretical conceptions and others who prove these conceptions empirically, use them, and develop them. Each of the "masters" possessing a certain theoretical conception is surrounded by a circle of "pot-boilers", who identify themselves as the supporters of that conception thus working it out or developing it. In this way they show that this conception explains even more than their "master" thought it did. From the point of view of an individual entering a given discipline it presents a certain objective state on which this individual has no influence. He may either expand the repertoire of theoretical conceptions functioning in this discipline or become another supporter of one of these conceptions. The first alternative indicates that the individual puts forward a new conceptualization of factors or that he proposes a new significance stratification of the known factors. If he does so he is wrong from the point of view of his further career (it concerns only typical cases). The reactions of the community of scientists towards the daredevil may vary significantly but by no means do they resemble the myth of "the group of competent and truth seeking researchers", with which this group identifies themselves. If we exclude certain rare exceptions, the daredevil is usually considered a man who cannot be taken seriously. This may end up in his civil death within the community with which he associates his social status, his ambitions and professional expectations. The reason is that from the point of view of the "masters", the appearance of a new theoretical alternative means that the young researchers of the future may direct their efforts not towards the development of their own conceptions but rather towards the development of the conception of their new rival. The appearance of a new theoretical alternative may also mean the emergence of a fronde in the rank and file of the masters' own supporters. Thus they spare no efforts in order to completely ignore the new theoretical proposal and depreciate its author.

The reaction of the group of scientists can hardly be considered surprising. After all, they are the ones who devote many years — sometimes their entire lives — to make specific researches based upon certain fundamental assumptions in which they believe so unwaveringly that sometimes they are not even fully aware of them. What else can their reaction be towards someone who is telling them that their hard work is wasted not because there exist certain deviations from the standards, but because these standards are wrong from the very foundations as they do not grasp the heart of matters, offering only apparent solutions? These people are only trying to defend themselves — they defend their status quo, their position in science, measured with the number of their supporters and the depth of their belief in the points of view offered by the "masters". And it is even easier for them to do it as they can — with no difficulty at all — find substantial arguments against the new, unavoidably schematic, and limited as a rule to the general outline of the first model proposition of the new theory. At the beginning the proponent of a new point of view is most often able to merely give a new repertoire of main factors. The establishment, not even speaking of an at least temporary completion, of secondary factors requires arduous researches, which cannot as a rule be carried out to a sufficient degree by the author in question. And so even if he overcomes the conspiracy of silence that his present "masters" create in response to his conception, he still becomes a target of a strong and often substantial criticism. Sticking to one's point means a long, sometimes many years long, battle — first against the conspiracy of silence, then against total criticism, and finally a battle for supporters, whose joint effort may lead a situation in which the new point of view wins for itself a position equal with the old ones.

This fate can only be accepted by very few individuals, which is not surprising. It is a rule, on the other hand, that a beginning scientist gives up saying something significant about the world on his own and he lets himself fall into one of the already existing points of view thus assuming the role of a "pot-boiler" of science. Most often he is not even aware of it as the existing ideology of science presents to him his accession in completely different terms: as a promotion to members of a group of competent researchers who work together on the development of a conception which has already been proved to be true. In this way the beginner loses the chance to say something significantly new. And he loses it irrevocably as the work within one of the existing theories overpowers his imagination with erudition and makes him lose the freshness of his view, which is so much necessary to notice the unknown theoretical structures in particular facts.

For those reasons the structure of the scientific community may be said to eliminate the opportunity of reaching beyond what is already known. The progress of science takes place in defiance of the natural mechanisms of science.

The structure of the scientific community has its material roots. The "masters" become what they are as soon as they obtain a possibility of recruiting their supporters. This, however, requires taking decision-making standpoints: access to publications, journals, etc., which make it possible to spread new conceptions and publish studies which develop them. In this way a "master" can find supporters and each work developing his point of view has — besides the cognitive dimension — also a doctrinal dimension: it wins new supporters for him. And since his position depends upon the number of supporters within the scientific community and since this number depends, in turn, upon the degree to which he can manage the means of scientific production within a given discipline, then his actions must be more and more directed towards the extending of the disposition of these means. Now, if the "master's" action is delayed, he will be anticipated by his rivals and he will lose — at least relatively — his position in favor of them. Competition forces one to care about the material roots of the scientific life even if it is not part of the intentions and predilections of a theoretician struggling for his position within the scientific community. The very same competition forces one to be intolerant in relation to new points of view and to treat any new young scientist as a potential "pot-boiler" able only to multiply the contributions to his "master's" own orientation. Therefore, similar to the economy, where competition is responsible for the spreading the motive of profit among the proprietors, and politics, where it is responsible for the spreading the motive of ruling, competition in science (competition among the "masters") is responsible for the motive of the multiplication of influences upon the minds of others becoming the main purpose of the masters' actions, putting the goal of a purely cognitive nature to the background.

The picture presented above is based upon every-day experience of scientists rather than upon any systematic empirical data. If we take the risk of accepting its accuracy, the structure of the scientific community in any scientific discipline may be presented in the following way:

1. Within a scientific community there exists a minority which controls the *means of scientific production*, i.e. the one which effectively makes decisions concerning the purposes for which these means are to be used. Members of this minority — let us call them the *priests* of a given science

— put forward certain alternative theories and use the remaining members of the community — constituting its vast majority — for the development of these theories (justifying them, using them to explain new phenomena, concretizing them). In this way a set of *pot-boilers* is attributed to each of the priests of a given science.

2. Each of the priests aims at extending the *range of his intellectual power*, i.e. at expanding the community of his pot-boilers and at strengthening their belief in the theory offered by him. Whoever of priests is delayed in doing this becomes eliminated by the competition from the group of disposers of the means of scientific production in a given discipline.

3. This produces a situation in which a new beginner entering a given science has very small chances of entering the category of disposers of the means of production characteristic of that discipline. For this reason he most often assumes the role of a pot-boiler of one of the existing theoretical alternatives. Thus he usually gives up developing his own new theoretical ideas in favor of elaborating to perfection the already existing ideas of others.

If it really happens so in science, then the answer to the question of whether a science is of an antagonistic nature is obvious. It is. However, it does not mean yet that the scientific community is a class system in the understanding presented above. In order to solve that question, we still have to elaborate some more on the relations between the priests and the pot-boilers, and particularly consider the question of what would be the nature of the influence exerted by the former upon the latter.

6. Intellectual domination

According to the idealizational approach to the nature of knowledge, the development of knowledge requires, above all, the multitude of theoretical alternatives, i.e. the occurrence of a possibly large number of various approaches to the given discipline — various conceptualizations of factors occuring in this discipline and various significance stratifications of these factors. The more there are such alternatives the higher the chance is for an appropriate conceptualization and stratification. Moreover, the multitude of different approaches allows for a utilization of different conceptions through paraphrase and a concretizational development of one's own conception (Wais (1979)). Thus the basic condition of theoretical activity in this approach is imagination — the ability to present a phenomenon as different from what it actually is but, at the same time, in such a way that certain characteristic features of it are preserved in the constructed model (the ideal type), and such features which make it possible to reduce the divergence between the idealized and the empirical world by means of concretizational operations.

Yet, the relations between the priest and the pot-boiler turn out to be depriving the pot-boiler of his creative imagination. The pot-boiler becomes the member of a scientific group, in which the leader is the priest-author of the basic model of a theory developed by the crew of pot-boilers recruited earlier. The existing theory assumes a certain conceptual apparatus, which distinguishes the factors of the given discipline, as well as a significance stratification of these factors. The condition for accepting a new scientist by such group is his adoption of two things — the way of conceptualization and the way of idealization of the given type of phenomena. The pot-boiler's unquestioned acceptance of these two things: the conceptual apparatus and the principles of distinguishing the basic model of a theory indicates that everything he can later do on his own is a completion and perfectualization of somebody else's theory. Thus he has to give up the use of his own creative imagination in favor of ancillary functions performed for someone who possesses a not neces-sarily intellectual advantage over him. The reason is that the priest is one of the disposers of the means of scientific production, who holds the keys opening the doors to publications, proper working conditions, not even mentioning the financial or substantial questions. And so the recruitment of pot-boilers in science is carried out in the same way as the recruitment of workers carried out by the capitalist. The only difference is that the worker sells his labour power in return for non-equivalent pay, i.e. in return for the pay burdened with economical exploitation, while a scientific pot-boiler sells his intellectual power in return for a non-equivalent position within a scientific group, i.e. a position burdened with intellectual inequality since it assumes the intellectual domination of a priest over the pot-boiler.

One could say that *A dominates intellectually* over *B* relatively to a certain class of beliefs if a sufficient and necessary condition for the acceptance of each of these beliefs by person *B* is his awareness that they are the beliefs of *A*. The minimum of the intellectual domination of a priest over a pot-boiler is demonstrated by the fact that the pot-boiler accepts the principles of conceptualization of the priest's theory and the laws of that theory comprising its basic model just because they come from his "master". As long as the pot-boiler remains what he is, the way of conceptualization of phenomena and the way of significance hierarchi-zation of factors expected by the priest's theory remain beyond discus-sion. On the other hand, the maximum of that domination consists in the fact that the priest requires the pot-boiler to accept all theoretical models comprising the theory which he, the priest, proclaims and just because of that. In this case the pot-boiler cannot even question the priest's establishings from the derivative models. He must accept not only the master's view on the nature of the analyzed phenomena but also his beliefs concerning the secondary circumstances. The range of the freedom of beliefs which is left to him by the "master" does not reach in such a case beyond purely factographic matters and the work which the pot-boiler can do is actually limited to finding facts confirming the accepted theory. This extreme case of the domination of a priest over a pot-boiler could be described as intellectual incapacitation.

And so it is possible to establish the essence of the peculiar social inequality between the priest of science and the pot-boiler of science. This essence consists in the fact that the former deprives the latter of the possibility to work creatively and he forces the latter to follow his own ideas performing ancillary functions. The priest forces the pot-boiler because the fact that the material conditions of the pot-boiler's work are at the hands of the priest puts the pot-boiler in a situation which is as much a situation of coercion as the one of a worker, who does not possess or control the material conditions of his work. The oppression charac-teristic of scientific activity consists in the fact that the disposers of the means of scientific production intellectually dominate over those who have no access to those means.

7. The idealized model of a scientific community: The scientific community as a class system

The answer to the question whether science (or, more precisely, a scientific community) is a class system is, for the above reasons, positive — the priests of science constitute a subclass (or, in other words, a faction) of the class of priests of culture, i.e. the class of disposers of all means of spiritual production. Such a qualification makes it possible to explain many commonly known phenomena, which are often (dis-) interpreted on the basis of current ideology.

Above all, it becomes obvious why the scientific community is so strongly hierarchical. Just like all the other ones this hierarchy is to consolidate the position of the strongest: to make the access to decision-making positions as difficult for the beginners as possible; to prolong the scientific career of a beginner in time, so that when he already reaches the highest positions he is only capable of subordinating himself to the already existing rules; to eliminate, on the subsequent qualificational stages, those who are not submissive and eager to violate the holy order: to bring the rebels within the framework of the system by giving them a certain, constantly growing on particular stages of their career, intel-lectual power over the beginning pot-boilers. Similar to the hierarchy of power, the hierarchy of science is an institutional superstructure reaching over the material monopoly — in the case of science it means the monopoly of disposing and controlling the means of knowledge productio. The "equality of scientists in relation to the truth" is a myth.

Further on, it becomes clear why every new conception is met with a conspiracy of silence and why it has to force its way to the surface with so much difficulty. The reason is that every new theoretical proposition endangers the interests of all those who had put themselves in support of the old one and who — by spreading the old one — had already won a certain position in the hierarchy that begins with pure priests and ends with pure pot-boilers. Cognitive arguments, which are drawn against the new proposal, are not motives for a decision to stick to the old theory but ideological rationalizations of the real motive: to preserve the intellectual influence which secures the domination of the old point of view. Empirical regularity, noticed by Max Planck, which is presented below can hardly be surprising:

"A new scientific truth does not triumph because it somehow manages to convince the opponents and make them see the light, but because the opponents die out and a new generation of scientists who know it well matures" (after Kuhn (1968), p. 166).

And, in fact, it is not their understanding of the truth that they are defending but their positions, their intellectual domination over others, worked out on the basis of the old point of view.

Finally, it becomes clear in the proposed approach why the practice of actions of the priests of sciences is so drastically different from their declarations and from what they say about science especially to people from the outside. They say that science constitutes probably the only free and true market, where certain opinions are supported by arguments rather than force. And yet all they do is aimed at accumulating power: the position of an editor of a book series or journal, or at least a lecturer in a publishing house or member of the editorial board. A typical professor is ready to spend any amount of time and effort to gain the highest possible influence upon what is published in large edition. It is the priests' influences that are the subject of maximization, that constitute the "function-criterion" which is realized by a typical scientist. Everything else: the real cognitive results, the income, etc. constitutes only "limiting conditions", which must be satisfied to an appropriate degree but which are not realized without any limit. On the other hand, what the priests say about it differs significantly because of its nature from the practice of their actions: they preach an ideology which — similar to the ideologies of private property or power — turns our attention from the reality by creating an illusion that conceals their interests. And it is usually they themselves who believe in those illusions, though the ideology of "pure science" is, above all, to serve the indoctrination of young researchers entering the scientific community.

8. Paradigm and a scientific revolution: A class approach

Let us now consider an idealized model of a scientific community, lacking any external influences. It constitutes a class system divided into a (sub)class of the priests of science and a (sub)class of pot-boilers. Let us assume that in the starting point each of the priests professes a different theory and, since he possesses certain means of knowledge production, he is surrounded by a circle of supporters.

At this point a competition between particular scientific clans (combined of a priest and a school of supporters) occur. The stake is the actual influence, that is gaining more and more powerful means of indoctrination encompassing a constantly larger part of the scientific community. Those that are weaker are eliminated together with their theories because they are less spread being surrounded by a smaller number of supporters. Seeing the position of their point of view get weaker, the pot-boilers join the one who is stronger. Naturally, in the competition between the clans there occurs a cognitive argumentation, which performs the role of an ideological ornament, though, in certain exceptional cases, it constitutes the actual motives of a possible change in the scientific orientation. Cognitive competition follows a general pattern of any competition: the strong become even stronger, while the weak become even weaker. The stake in the game is not the pot-boilers, who are already followers and supporters of a certain idealizational theory — the education of a scientist, the way of teaching him the principles of factors conceptualization, significance stratification, criteria of significance, etc. may sudden theoretical changes very difficult — but new students, who are only beginners in the given science. Those, however, are the ones who join the largest and most powerful clans, that is the groups of most influential priests.

The end of competition is similar to the end of any free market: it means self-suppression and therefore generating of a cognitive monopoly. We may assume it within the framework of our idealized model of an isolated science that in the end the competitors are eliminated or they withdraw themselves. In this way the entire scientific community falls under the power of one point of view, or one *paradigm*.[1]

A paradigm is a certain theoretical orientation, or a pair (the principles of essential stratification, idealizational theory), the supporters of which have won in the competition with other theoretical orientations. They won because they were powerful and not necessarily because the truth was on their side, which means that they managed to take possession of all the means of knowledge production, eliminating their rivals from the market of scientific ideas. From this moment the entire scientific community works for the given paradigm according to a specific division of labour. Problems philosophically fundamental (ontological and epistemological problems lying at the basis of the principles of essential stratification of the paradigm) are undertaken exclusively by the priests of the highest rank and almost exclusively in an ideological way. The philosophy lying at the basis of the paradigm can only be a sacrum, which cannot be discussed seriously but which must be popularized as a dogma staying beyond discussion. As time passes, also the basic, most idealized model of the paradigm undergoes sacralization: the problems connected with it may be undertaken only by the elite of the priests. The priests' every day work is connected with searching for subsequent concretizations of the model according to the principle of dialectic correspondence (Nowakowa (1975)): finding deviations from the basic model, identification of factors treated from the beginning as secondary ones and responsible for the deviations, the concretization of this model from the point of view of these factors carried out in order to eliminate these deviations. The pot-boilers' every day work is even more commonly. They cannot even correct the paradigm — an inappropriate attitude towards the deviation could lead not to the correction but a falsification or, in other words, a blow into the spiritual basis which is the main support for the priests' interests. The only thing the priests allow the pot-boilers to do is to search for new uses of the paradigm, that is developing consequences from the assumptions of the paradigm treated as an unquestionable whole.

The scientific community is structuralized according to the position which can be gained through the work on the paradigm. The more interesting are the uses of the paradigm which one discovers and the more significant corrections solving difficult problems he obtains, the higher is position he gains in the scientific community. In this way a pot-boiler may change his class qualification: gaining access to the means of knowledge production he can become a priest. However, in the course of the development of the paradigm, such a social promotion is more and more difficult to be obtained. Growing sacralization of the paradigm (first including only the stratificational principles, then the basic model, and finally all the subsequent derivative models) makes the correctional area — and then, in the ideal case, also the area of applications — disappear. Thus not only the master-piece of a less or more mythical creator of the paradigm becomes unquestionable but also the correctional discoverings of the constantly growing number of "grand masters" of the paradigm. In the result, however, it interferes with the interests of young scholars by closing the channels of the social mobility. New scholars do not consider joining the paradigm as their interest any more. The scientific community comes into a revolutionary situation.

Together with this situation a change in the attitude towards the revolutionaries occurs. Nonconformists opposing the constructional principles of the paradigm and proposing new alternative theoretical schemes have always existed within the scientific community. As long as the paradigm was in the state of development, the revolutionaries were considered "unreasonable people" and were pushed aside. They did not have any serious chances to gain supporters for the pot-boilers tended to get their promotion in the easy way — by staying in accordance rather than the opposition to the paradigm. Blocking the opportunity of promotion within the paradigm changes the situation quite radically. Developing alternative theoretical orientations gain supporters among the new scholars, who see a better chance for themselves within the new orientation than within the paradigm. The scientific community re-enters the period of competition between a certain number of theoretical orientations. The old paradigm is only one of the alternatives, faithful to which are only those who are not able to obtain a higher position within any of the alternative orientations, that is the clan of priests. As a result of the mechanisms of competition described above one of the new points of view wins and becomes a new paradigm of the scientific community. And so on.

Therefore, the development of an idealized, isolated from any external influences, scientific community occurs according to the following scheme: competition — paradigm — revolution — competition — paradigm. This model is reproduced — with our simplifying assumptions — without any limits. Moreover, as long as we consider this problem from the point of view of internal mechanisms of the scientific community, there is no guarantee that the given paradigm will be cognitively better than the preceding paradigm. The class nature of science does not, by any way, secure cognitive progress.

9. The source of the progress of science

The above does not mean that there is no progress in science. If the progress of science is to be measured by essential truthfulness, subsequent natural theories recognize a constantly larger number of significant factors and, beginning with a certain stage of the development of knowledge, they recognize even the main factors — a fact, which has already been pointed out (Nowakowa (1977), Nowak (1980), Magala-Nowak (1985)). And so the process of passing from relative truths to the absolute truth occurs at least in the natural sciences; I do rather not talk about the humanities here. How can this be possibly explained if the internal mechanisms of science by no means lead to it?

There is only one possible explanation: science owes its cognitive progress not to itself but to external factors. And it is a fact: any state requires science to produce constantly more effective weapons and constantly more effective means of control over the society; private property requires science to produce constantly more efficient or more profit-giving means of production; the priests of "mass culture" require science to produce constantly more effective systems of persuasion. And so science has to produce effective devices of an economical, political and spiritual control over people. Science is a servant of the classes of oppressors, single or multiplied. For that reason it is forced to produce a constantly deeper and more complete truth about everything which could make the mighty of this world even more powerful.[2]

10. A conclusion

Let us now put together the most important points of the approach presented here.

1. It is necessary to distinguish between the epistemological and the social dimension of science as well as, respectively, the methodology of knowledge and the sociology of science. Irrespective of what goes on in science, the theses of methodology are true or not; similar to logical tautologies that are true irrespective of whether there are people who structure their ways of thinking on the basis of them. On the other hand, the sociology of science is a social science which explains the statics and dynamics of the scientific community.

2. The scientific community is of a class character and is divided into the priests (disposers of the means of knowledge production) and the pot-boilers, upon whom the priests impose their point of view. The scientific community must then be analyzed — similar to the state or economic community — in antagonistic terms rather than the solidaristic ones.[3]

3. The scientific community develops according to the scheme: competition — paradigm — revolution — competition — paradigm, etc. The passing from one stage to another has, usually contrary to Kuhn's (1968) concept of a paradigm, not much in common with cognitive reasons and constitutes a result of social processes, in which cognitive argumentations perform, statistically speaking, the role of ideological ornamentation that rationalizes decisions motivated by the tendency to intensify the intellectual influence upon other members of scientific community.

4. The *scientific* community does not include any internal mechanism securing the progress of *knowledge*. It produces — at least in certain disciplines like, for example, the natural science — knowledge which is constantly more essentially true because it must secure constantly more effective technological solutions. The latter is required from the priests of science by the classes which rule politically, economically and/or

spiritually over the masses. For that reason, in those scientific disciplines which can be used technologically, subsequent paradigms acquire a constantly growing explanatory power.

Translated by Krzysztof Sawala

REFERENCES

1. T.S. Kuhn's conception of paradigm is based upon the assumption that cognitive motives are a crucial type of the determinants of cognitive actions. And so, according to Kuhn, a paradigm is a means of research standardization which makes cognitive progress possible: certain acceptable types of the modern scientific practice constitute a model from which a particular compact tradition of scientific research emerges. The formation of a paradigm and more specialized research that it allows is a symptom of maturity of the given scientific discipline (Kuhn (1968), p. 27, 28). The scientists' resistance to giving up a paradigm is cognitively rational: a strong resistance, especially of those who have subordinated their activity to the old tradition of institutional science, is an expression of the essence of scientific work. The source of this resistance lies unquestionably in the belief that the old paradigm will finally solve all its problems by itself and that nature can be put into the very drawers which this paradigm had prepared for it (op. cit., p. 167). Also the decision concerning the support for the new paradigm is based on cognitive grounds: the problem lies in the question of which of the paradigms is going to control the scientific analysis of the problems that could not be solved by any of the competitors. Anyone who chooses a given paradigm at an early stage of its development must believe that the new paradigm will win the future competition with many difficult problems, although all he knows is that the old one had failed several times (op. cit., p. 173).

2. In this way I change the view that it is the scientific competition that secures a constantly higher level of essential truthfulness in science (Nowak (1980)). Searching for the sources of the progress of science in the internal mechanisms of the scientific community resulted from the solidaristic approach to that community, which I used to accept before I worked out non-Marxian historical materialism.

3. Here is an example of the solidaristic treatment of scientific community: the scientific community consists of the specialists of a certain scientific discipline. To a much larger extent than in the case of a majority of other disciplines these specialists received similar education and similar professional initiation, read similar literature, and drew similar conclusions from it. The members of the scientific community are then considered by themselves as well as by others as those who are burdened with exclusive responsibility for pursuing common goals, including the education of their followers. Within these groups the mutual agreement and cooperation is almost complete, and the professional evaluation — almost unanimous (Kuhn (1977), p. 96).

LITERATURE

1. Kuhn T.S. (1968), *The Structure of Scientific Revolutions* (the Polish translation), Warsaw
2. Kuhn T.S. (1977), "A Post-Script of 1969 to The Structure of Scientific Revolutions" (the Polish translation), *Zagadnienia Naukoznawstwa*, vol. XIII, no. 1
3. Magala S., Nowak L. (1985), "The Problem of Historicity of Cognition in the

Idealizational Conception of Science", in: (ed) J. Brzeziński, *Consciousness: Methodological and Psychological Approaches, Poznań Studies in the Philosophy of the Sciences and the Humanities*, vol. 8

4. Nowak L (1971), *Foundations of Marxian Methodology of Science* (in Polish), Warsaw

5. Nowak L. (1979), *Foundations of the Theory of Historical Process* (in Polish, samizdat), Poznań, 2 vols.

6. Nowak L. (1980), *The Structure of Idealization. Towards a Systematic Interpretation of the Marxian Idea of Science*, Reidel, Dordrecht/Boston/London

7. Nowak L. (1983), *Property and Power. Towards a non-Marxian Historical Materialism*, Reidel, Dordrecht/Boston/Lancaster

8. Nowakowa I. (1975), "Idealization and the Problem of Correspondence", *Poznań Studies in the Philosophy of the Sciences and the Humanities*, vol. 1, no. 1

9. Nowakowa I. (1977), *Problems of the Theory of Truth in Marxist Philosophy* (in Polish), Poznań

10. Wajs J. (1979), "The Rule of Tolerance in the Idealizational Conception of Science" (in Polish), *Poznańskie Studia z Filozofii Nauki*, vol. 4, Warsaw/Poznań

Julius Sensat, Jr./Milwaukee

RECASTING MARXISM: HABERMAS'S PROPOSALS

I

The long period of economic prosperity experienced in the developed capitalist world following World War II convinced many people that state intervention in the economy had eliminated capitalism's inherent tendency toward economic crisis. It seemed as though capitalism had undergone a "mutation" which eliminated its deficencies as a vehicle for the development of the productive forces. The classical Marxist critique of capitalism thus seemed no longer capable of being substantiated. Moreover, it appeared that the bourgeoisie was no longer producing its own gravediggers. Various economic, psychological and cultural forces seemed to be producing a working class willing to identify with the economic structure, provided it would "deliver the goods", and active involvement of the state in the economy seemed to insure that this demand would be met. Ironically, such acquiescence to the requirements of capital seemed to be called for by the tenet of historical materialism that classes gain and maintain ascendency on the basis of their ability to preside over the development of the productive forces. A historical materialist should expect capitalists to rule, if capital places no fetters on the productive forces.

Frankfurt-School theorists tended to endorse this analysis, but they did not, as materialism seemed to dictate, rejoice in the fact that the productive forces had finally entered into a stage of unchecked development. Something seemed terribly wrong with modern capitalism, and if this intuition could be articulated within an effective materialist critique, then something must be wrong with materialism. This "something" might have to do either with materialism's standards of historical progress or with its claims about the determinants of historical change. The Frankfurt School tended to adopt the former diagnosis. Development of the productive forces was firmly in charge of history, but this triumph of "technical reason" was, contrary to the claims of the Enlightenment in general and Marx in particular, a perversion rather than a realization of human values. But whether one rejected historical materialism's empirical or its normative claims, the upshot seemed clear:

there was no materialist basis for expecting a social crisis which would issue in a fundamental and progressive transformation of society.

Habermas rejects this supposed implication. Hope for progressive change can rest on a materialist analysis of modern capitalism, he maintains, in spite of—or even, ironically, *because* of—the mutation brought about by state intervention in the economy. Modern capitalism is susceptible to crisis not because it places fatters on the forces of production—it does not[1]—but because the state's economic policies, though potentially effective, will not receive the political authorization they require. Such a crisis would not at bottom be an economic one, but rather a crisis of legitimation.

The distinction between an economic crisis and a legitimation crisis can be misleading, since every social crisis involves the withdrawal of legitimation in that people cease to regard existing social arrangements as justified. Habermas's aim is to distinguish between two possible sources of social crisis: (i) the inability of capital to mobilize and develop the productive forces, and (ii) the inability of the advanced capitalist state to satisfy simultaneously the requirements of capital accumulation and those of democratic determination of resource allocation and product distribution. In asserting (ii) as a source of crisis which is independent of (i), Habermas is denying the tenet of historical materialism that social crises arise only on the basis of a contradiction between the forces and relations of production. He concedes that his view is inconsistent with historical materialism as classically formulated, but he argues that the classical formulation is deficient. Suitably recast, historical materialism allows the sort of crisis-susceptability he wants to affirm.

The recasting proceeds roughly along these lines: Human beings not only produce; they also communicate with each other. Communication, like production, involves the exercise of certain abilities. Instead of calling these communicative forces—an appellation which would make more explicit the parallel he wants to draw—Habermas groups them together under the rubric "communicative competence." Like the productive forces, communicative competence can develop in history. A given form of society can enter into a crisis even when it is an adequate vehicle for the use and development of the productive forces, provided it precludes realization of the highest form of communication within reach of people's communicative competence. In other words, a "contradiction of material life" can take the form of a fattering by society either of productive forces or of communicative competence. It is the latter sort of contradiction, Habermas wants to claim, which characterizes modern capitalism. More and more people have the competence to determine "discursively"—i.e., via discussion guided solely by the force of the better argument—the needs which society ought to direct its resources toward satisfying, whereas capitalism prohibits such discussion, since it turns essentially on the private appropriation and disposal of society's resources.

Habermas's work constitutes an important contribution to twentieth century Marxist discussion. His investigations hold the promise of providing a deeper understanding of the nature of class struggle and the form it takes in modern times. They may help to resolve certain issues crying out for treatment within Marxism— issues concerning the nature of historical progress, the justifiability of moral claims and their role in a materialist critique of capitalism, and the conditions for the reproduction and eventual overthrow of patriarchal as well as capitalist social relations. Section IV of this essay provides brief elaborations of these themes.

However, the importance of Habermas's contribution does not derive from its being the most appropriate response to the apparent mutation brought about by state intervention in the economy. The context in which Habermas began working out his ideas was one characterized by an overestimation of the autonomy of the state vis á vis capital. Habermas's work does not expose this overestimation but rather is itself infected by it. This defect should not be left unremedied in a period in which it is becoming increasingly clear that the forms of state intervention which came to fruition during the postwar years were capable of insuring only temporary — if phenomenal— capital accumulation. Habermas should not be rejected by Marxists as simply one more social theorist taken in by the illusion of the possibility of permanent economic prosperity on a state-managed capitalist basis. Nor, for that matter, should he be viewed as simply one more "revisionist" who overestimates the democratic potential of capitalist political institutions. His investigations have a significance which can only be obscured by their ties to these misperceptions. Consequently, most of the following consists of criticism, the primary aim of which is to break these ties. Section II argues that it is harder than Habermas thinks to establish that the state is no longer superstructural in the classical sense, and then discusses some barriers to genuine democracy which Habermas apparently does not take seriously enough. Section III attacks the claim that advanced capitalism does not fetter the productive forces. Habermas is concerned to reject a supposed implication of this claim, namely, that there are no longer materialist grounds for expecting a progressive social transformation, but he endorses the claim itself. As a result, he rejects the classical Marxist critique of capitalism in favor of the argument that the requirements of capital accumulation conflict with those of democratic resource allocation and product distribution. There are no grounds for such an exclusionary approach. The fundamental claim of Marx's critique of political economy, that every form of capitalism is inherently fated to become dysfunctional for the development of the productive forces, is sufficiently plausible and important that ignore it is to risk delusion about the future as well as abandonment of a potentially decisive critique of capitalism.

II

According to Habermas, the case for the claim that a legitimation crisis is to be expected in modern capitalism rests on the following four premises:

(1) The economic activities of the modern state must be legitimated through political institutions democratic in form.

This is supposed to follow from the fact that the rise of capitalism brought with it an evolutionary advance in normative structures to the universalistic level (Habermas, 1973a, p. 54; 1975, p. 36).

(2) Extensive participation in these institutions would result in a withholding of legitimation for state activity aimed at securing capital accumulation.

Political institutions democratic in form presented no threat to capitalist organization of the economy so long as the ideology of fair exchange effectively restricted political activity to securing the boundary conditions of the market economy. However, the assumption by government of the role of responsible planning authority publicly acknowledges the insufficiency of the market to achieve social justice. Consequently, in advanced capitalism the allocation of resources and the distribution of the social product become in principle matters of public concern. Under these circumstances the existence of formally democratic institutions presents severe problems for the continued existence of capitalism. For these institutions provide a means by which people can raise political demands for the accommodation of needs whose satisfaction was formerly the exclusive province of the market. The more people who avail themselves of this option, the harder it becomes for the state to pursue policies which are both economically adequate and socially acceptable. Only a high level of political abstinence can prevent such a bind form arising. This abstinence cannot be enforced directly by legal means, in virtue of (1); rather, it must be secured in the socio-cultural sphere through privatistic orientations away from politics and toward both status achievement through competitive career advancement and family-centered consumption and leisure. However:

(3) Conditions for the reproduction of the motivational syndromes of privatism whose function is to insure political abstinence on the part of the general populace are being eroded (Habermas, 1973a, 1975, pt. II, ch. 7).

The above propositions support the thesis that the problem of legitimation which faces the modern state cannot be *solved*. However, it is conceivable that the problem could be *eliminated*, through the adoption of a mode of socialization which proceeds by way of "instinct-like self-stabilization" (Habermas, 1968, p. 97; 1970, p. 118) of a social structure whose identity is formed independently of interpersonal discussion. Habermas sees a practical role here for his theoretical work: he appeals to the opposition to such a society which he sees as implicit in every communicative act. Provided that such opposition can achieve and maintain political efficacy, then

(4) the problem of legitimation cannot be eliminated altogether, and capitalism is headed for legitimation crisis.

The following remarks will focus on the defensibility of (2). First, I shall explain how (2) presupposes that the modern state is not superstructural vis á vis capital. Second, I shall criticize Habermas's main argument for this presupposition. And finally, I shall discuss some conditions which would falsify (2) and the possibility of which Habermas does not take seriously enough.

In the classical version of historical materialism[2], an institution is superstructural with respect to a certain type of economic structure (e.g., capitalism) if it obtains because it functions to sustain a structure of that type which is an appropriate vehicle for use and further development of the productive forces. Habermas needs to deny that the modern state is superstructural with respect to capitalism in this classical sense, since (2) entails that political institutions would permit a withholding of legitimation even of state policies which would be economically effective, i.e., would insure an acceptable level of use and development of the productive forces through capital accumulation. According to the classical view, political institutions can never play such a destabilizing role; they can be an agent of fundamental transformation of the economic structure only when the latter acts as a fetter on the productive forces.

Habermas's central argument against the claim that the modern state is superstructural with respect to capitalism does not work. For it turns on an entailment claim which will not stand up to scrutiny, namely that the non-superstructural character of the state follows from the fact that the relations of production have a political character in modern capitalism. The following is a fairly recent version of the argument (Habermas, 1973a, p. 76; 1975, p. 52):

Government activity now pursues the declared goal of steering the system so as to avoid crises, and consequently the class relationship has lost its unpolitical form. Hence class structure *must* be maintained in struggles over the administratively mediated distribution of increases in the social product, and therefore the class structure now *can* be directly affected by political disputes as well.

It might seem as though this inference is straightforwardly valid. For in order for the economy to have the explanatory primacy with respect to the state that the relation of base to superstructure demands, the state must be separated from the economy, and if the relations of production are political, then, it would seem, this condition is not satisfied.

It is true that the state cannot be superstructural with respect to the economy unless society can be differentiated into a legal and political structure, on the one hand, and an economic structure on the other. However, this differentiability is guaranteed by the fact that legal and political relations are different in kind from economic relations. The latter are *power* relations which govern how the productive forces are used, whereas the former, though they often sustain or reinforce such power relations, are relations specifying *rights and obligations* which people have concerning the use of society's resources.[3]

This difference in kind is not eliminated by the new economic functions of the state in advanced capitalism. The sense in which there has been a "politicization" of production relations has to do with *how* economic relations are sustained by the legal and political structures and not with the *nature* of those relations themselves (i.e., whether they have been transformed from power relations into legal and political ones). The question is really whether disposal of productive resources is authorized not only by legal recognition of private contracts but through governmental directives and policies as well. If so, then the state is directly involved in the economic process; this is the legitimate meaning of the claim that the relations of production are political. But the truth of this claim in no way precludes legal and political institutions from being superstructural vis á vis capitalism. To repeat: legal relations and forms of state are superstructural with respect to a certain type of economic structure if they obtain because they sustain a structure of that type which is an appropriate vehicle for use and further development of the productive forces. It may be that only a state directly involved in the economic process can sustain a form of capitalism which is such a vehicle, given the present state of development of the productive forces. And if the modern state exists *because* of this functionality, then it retains superstructural status.

In drawing the inference criticized above, Habermas is falsely identifying the superstructure-base relation with the state-civil society relation. For him, the state stands to the economy as superstructure to base exactly when the classic separation between the state and civil society, characteristic of the early phase of capitalist development, obtains. But since such a separation can exist only in predominantly commodity-producing societies, where production is immediately private and only indirectly social, this reading of the superstructure-base relation would render the claim of classical historical materialism, that the economic structure is "the real foundation, on which arises a legal and political superstructure", (Marx, 1859, p. 20) *automatically* false of non-commodity producing societies. Marx clearly meant to assert this claim with respect to all historical social formations, including those in which he was aware that state and civil society could not be clearly separated. On Habermas's reading, in doing so he would be making an elementary logical blunder. The claim may not be true of all societies, but on my reading whether or not it is true is always an empirical question rather than a matter of logic.

Let us grant that discursive resolution of issues concerning resource allocation and product distribution is incompatible with acquiescence to the requirements of capital accumulation. It does not follow that extensive participation in today's political institutions would issue in non-acquiescence. Habermas tends, I think, to overestimate the disspelling effect which state intervention in the economy has on the mystifications arising out of the capitalist mode of production. He focuses on the ideology of equivalent exchange — the view that market transactions are just because they involve the exchange of equivalents — and asserts that the market-tampering activities of the state involve an official recognition of its falsity. But even if this be granted, the notion is still widespread that the way to have economic prosperity is to provide "jobs", and the way to provide jobs is to "create a favorable climate for investment". This notion is deep-rooted; it is an offshoot of a fetishism inherent to capitalism, the false appearance of the forces of production as capital by nature. Capital does not appear to be productive in virtue of its embodiment in the forces of production; rather the latter appear to have their productive character only as incorporations of capital.[4]

Moreover, even if the spell of these illusions were broken and people by and large did become communicatively competent enough to engage in discursive determination of the proper disposal of society's resources, it is not at all clear that it would take place. The barriers to what Habermas calls "substantive democracy" are not merely ideological; they are also *economic*. People come into the political arena with different resources and powers, derived from their different economic positions. Formal democracy, with its equality of legal rights of participation, does not preclude systematic inequality of power across social classes to occupy, or to influence those who occupy, official decision-making positions and to determine both what issues get addressed and how they are defined. This problem is exacerbated by political structures which diminish the sensitivity of official political channels to popular will (Edwards and Reich, 1978). For example, in the United States, many issues of substantial importance are decided neither by national referenda nor by elected bodies, but by appointed officials. Formal democracy is preserved in the sense that these officials are ultimately appointed by elected officials, but popular accountability is very indirect and attenuated. Moreover, the lack of a system of proportional representation in the United States hinders democratic will-formation by inhibiting third-party efforts and channeling electoral activity through the two major parties, where capital has a major foothold. This does not mean that socialists should abandon the political arena as a focus of class struggle; on the contrary, they should work to increase the responsiveness of political institutions to working-class interests, but they should do so with a realistic understanding of the forces involved. It may be that such a battle can be won only on the basis of a solid case against capital's viability as a form of development of the productive forces.

III

To assess adequately what I have characterized as Habermas's response to the pessimism of the Frankfurt School, we must examine the assumption which precipitated their reorientation toward historical materialism: that modern capitalism does not fetter the productive forces. Habermas questions the pessimism which this hypothesis evoked, but he does not question the hypothesis itself. In fact, he offers both negative and positive support for it. On the negative side, he argues that Marx's theory of capital development has been rendered inapplicable as a crisis predictor by the politicization of the relations of production. On the positive side, he offers the rudiments of a theory of social evolution and an analysis of modern capitalism in terms of the basic concepts of the theory; according to this analysis, productive forces are not fettered in modern capitalism. We shall discuss each of these in turn.

Habermas has frequently argued for the inadequacy of Marx's theory of capital development on the grounds that state and economy are no longer related to each other as superstructure to base (1968, pp. 75-6; 1970, p. 101):

The institutional framework of society was repoliticized. It no longer coincides immediately with the relations of production, i.e., with an order of private law that secures capitalist economic activity and the corresponding general guarantees of order provided by the bourgeois state. But this means a change in the relation of the economy to the political system: politics is no longer *only* a phenomenon of the superstructure. If society no longer "autonomously" perpetuates itself through self-regulation as a sphere preceding and lying at the basis of the state — and its ability to do so was the really novel feature of the capitalist mode of production — then society and the state are no longer in the relation that Marxian theory had defined as that of base and superstructure. Then, however, a critical theory of society can no longer be constructed in the exclusive form of a critique of political economy. A point of view that methodically isolates the economic laws of motion of society can claim to grasp the overall structure of social life in its essential categories only as long as politics depends on the economic base. It becomes inapplicable when the base has to be comprehended as in itself a function of governmental activity and political conflicts.

If the state were no longer superstructural with respect to capital in the sense of the last section, then crises could in principle arise from sources other than contradictions between the forces and relations of production, and crisis theory developed along classical Marxian lines would not yield a complete specification of society's developmental dynamics. However, as we have seen, that such a change has occurred cannot be established merely by citing the politicization of the relations of production, as is done in the above passage. In fact, we see clearly in this passage the mistaken identification of relations of production with legal relations and the consequent confusion of the superstructure-base relation with the state-civil society relation.

Moreover, Habermas wants to make a stronger claim of inadequacy for Marx's theory than simply that it specifies only *economic* crisis tendencies when other sorts are possible as well. He denies Marx's assertion that economic crises are inevitable. State intervention, he thinks, has made this assertion unsupportable. And this stronger claim cannot be deduced, it seems to me, merely from the supposed non-superstructural character of the state.

However, Habermas does not leave the discussion at this level: he gives a specific argument for the stronger claim. According to him, governmental organization of higher education and of "research and development" has provided a source of surplus value not countenanced by Marx's theory and has thereby rendered inapplicable the law of the tendency of the profit rate to fall.

In considering this argument, it seems important to distinguish between an economic crisis and a periodic business-cycle recession. While the term "crisis" has been used to denote the latter, it can be misleading to do so, since in this kind of period of economic slowdown normal economic activity within the prevailing institutional context is sufficient to restore prosperity. In the history of capitalist formation, however there have been periods of major economic instability and stagnation which strained the entire social fabric. Major changes in social structure were needed to regenerate the process of capital accumulation. The term "crisis" seems more appropriately reserved for these economic conditions.[5]

Organized political intervention in the economy has always been one of the ways in which resolution of economic crisis through social restructuring has been brought about. In this respect Habermas's distinction between liberal capitalism, in which the state does not intervene in the "functional gaps of the market", and advanced capitalism, in which it does, is inaccurate. Whenever the socio-economic structure comprising the particular form of capitalism in existence has become dysfunctional for continued accumulation in the sense that normal economic activity within the framework of that structure cannot restore accumulation, the state has intervened. Such intervention has been one of the primary means whereby a new "social structure of accumulation"—to use Gordon's apt phrase (1980)—has been established. It is true, however, that the state plays a more direct role *within* the functioning of the present social structure of accumulation than in previous ones; this is the phenomenon to which Habermas attaches crucial significance. The question we need to address is whether this new role of the state eliminates the inevitability of economic crises to which previous structures were subject.

In formulating his theory of the tendency of the profit rate to fall, Marx recognized the existence of certain "counteracting influences" which worked to raise the rate of profit. Increases in productivity can effect both an increase in the rate of surplus value and a cheapening of the elements of constant capital. To these upward influences, however, Marx accorded a strictly subordinate status; they might bring about temporary upward fluctuations in the rate of profit, but in the long run the downward tendency would win out.

Habermas accepts this claim of long-run inevitability for liberal capitalism, but rejects it for advanced capitalism. The reasoning is as follows: The counteracting influence of increases in productivity was limited because of the fortuitous nature of their source; they resulted from the utilization of inventions and information which were externally generated relative to the economic system.

The institutional pressure to augment the productivity of labor through the introduction of new technology has always existed under capitalism. But innovations depended on sporadic inventions which, while economically motivated, were still fortuitous in character (Habermas, 1970, p. 104).

In other words, in liberal capitalism there was no systematic generation of economically utilizable, productivity-increasing innovations.

The advanced capitalist state, however, functions so as to incorporate the labor of scientists, engineers, and teachers into the economic system itself.

Only with governmental organization of scientific-technical progress and a systematically managed expansion of the system of continuing education does the production of information, technologies, organizations, and qualifications become a component of the production process itself. Reflexive labor,that is, labor applied to itself with the aim of increasing the productivity of labor, could be regarded at first as a public good provided by nature. Today it is internalized in the economic cycle. (Habermas, 1973A, p. 81; 1975, p. 56).

This economic institutionalization of "indirectly productive labor" vitiates the inevitability of economic crisis asserted in the theory of the tendency of the profit rate to fall. Such crises may still occur, but they can no longer be systematically predicted. The structure of the system no longer guarantees the dominance of the falling tendency of the profit rate over its counteracting influences.

Regardless of the truth or falsity of these assertions, they reflect a different approach from that of Marx to the question of a tendency of the profit rate to fall, and indeed to crisis theory in general. According to Marx, capital has an immanent barrier; this means that it is a type of economic structure whose very nature precludes it from being a permanently viable form of development of the productive forces. It is in some respects analogous to a biological organism whose life-span is doomed to be finite because of its genetic make up, regardless of environmental conditions. To be sure, the course of any actual crisis is to be explained by reference to external as well as internal factors, just as the actual development of the organism is to be explained by reference to interaction between genes and environment. The food supply made available to the organism by its environment affects its actual course of development; similarly, the suitability of the climate for agriculture and the availability of natural resources affect the workings of the entire economic system. Marx, however, did not see the inevitability of crisis as resting on any particular extrasystemic conditions. This is why he said "the real barrier to capital is capital itself". (Marx, 1894, p. 250). In his view, capitalism is *inherently* fated to become dysfunctional.

The theory that the rate of profit has a tendency to fall is a particular articulation of this view. Marx repeatedly criticized attempts to found this tendency on factors external to capital. For example, he said of Ricardo, whose explanation referred to decreasing productivity in agriculture due to cultivation of less fertile lands in the course of capital accumulation, that "he flees from economics to seek refuge in organic chemistry" (Marx, 1857-8, p. 754). Though Marx on one occasion reverted to a Ricardian type of argument himself (Marx, 1862-3, p. 368), his central aim was to demonstrate a falling tendency of the rate of profit rooted in factors internal to the capital relation itself. Thus while Habermas tries to ground the tendency in an environmental insufficiency, namely a lack of sufficient input of productivity increasing innovations, Marx thought he had located a barrier to accumulation which would arise regardless of — indeed in some sense because of — increases in productivity.

The rate of profit does not fall because labor becomes less productive, but because it becomes more productive. Both the rise in the rate of surplus value and the fall in the rate of profit are but specific forms through which growing productivity of labor is expressed under capitalism (Marx, 1894, p. 240).

If this hypothesis by Marx is correct, then institutionalization of the production of productivity-increasing innovations is not sufficient to offset the tendency in the manner Habermas suggests.

It might be objected that the foregoing reasoning relies on a distortion of Habermas's characterization of the transition from liberal capitalism to advanced capitalism. The transition is not one where the same economic system comes to confront a new, more favorable environment; rather, what is involved is a fundamental change in economic structure, one which alters the boundaries between system and environment. The appropriate analogy is that of a person who tries to remedy heart malfunction not by moving to healthier surroundings but rather by surgical installation of a pacemaking device. Such a structural change could well render certain dynamical principles inapplicable.

Is advanced capitalism capitalism? If so, then the above objection is unsound. The theory that capital has an immanent barrier is the view not that this or that particular type of capitalism is inherently fated to become dysfunctional but rather that capital *per se*, as a general type of economic structure, is so fated. Consequently, either the theory is true of all variants of capitalism or it is true of none. Regardless, then, of important structural differences between them, if the theory is true of liberal capitalism it is true of advanced capitalism, provided that the latter is indeed a form of capitalism.

Of course, such a theory may be quite difficult to establish; it is certainly not obviously true. Support for it requires arguments which appeal neither to exogenously determined insufficiencies in the system environment nor to peculiarities of a particular variant of capitalism. No arguments of this kind have been offered above. The focus of the foregoing discussion has been more limited. The restrictions on verifying Marx's theory entail corresponding restrictions on refuting it: it cannot be dismissed by appealing to peculiarities of advanced capitalism. Habermas's criticisms of the theory miss their mark and consequently do not provide the alleged negative support for his view that modern capitalism does not fetter the productive forces.

The foregoing criticism would be of dubious relevance if the theory that capital has an immanent barrier were clearly unsupportable. It is true that there is a gap in Marx's argumentation for the law of the tendency of the profit rate to fall. He offered arguments which, suitably refined, show that an organic composition of capital which is secularly rising (without limit) will eventually depress the rate of profit (regardless of accompanying increases in the rate of surplus value). But he did not provide sufficient warrant for the key claim that such a secular rise is inherent to the process of capital accumulation. However, this issue is by no means a dead horse; debate continues, at a high theoretical level[6].

Moreover, it is quite possible that further investigation will reveal that the consequences of the absence of a secular rise in the organic composition of capital are just as crisis-ridden as those of a secular rise. If this is the case, then the theory that capital has an immanent barrier is correct, though Marx's particular version of this theory is not[7].

I shall close this section by discussing two ways in which Habermas's incipient theory of social evolution wrongly biases his view toward the claim that advanced capitalism places no fetters on the productive forces. The first has to do with a particular hypothesis about social evolution which he endorses; the second stems from inadequacies in the articulation of basic concepts of the theory.

At one point in sketching the rudiments of his theory, Habermas asserts that development of the productive forces always increases social stability, provided that it does not precipitate destabilizing advances in communicative competence (Habermas, 1973a, p. 24; 1975, p. 12). His reasoning is presumably that productive-force development, by increasing society's power over nature, works to make the environment of the social system more manageable and thereby works to increase stability. If defensible, this hypothesis would certainly lend weight to Habermas's claim that advanced capitalism does not fetter the productive forces, for it rules out much that would count as fettering from occuring in *any* society. It apparently denies, for example, that a growth in productive capacity could occur in a society without that society's being able to use that capacity, because of structurally determined inefficiencies of resource utilization in the economy. No increases in power over nature would seem to accompany such a development. The hypothesis also does not countenance the possibility that future development of new, superior technology could be rendered technically feasible by growth of the productive forces while at the same time remaining socially infeasible because precluded by the relations of production. Such preclusion might well increase social instability, apart from any effects it might have on communicative competence.

But surely such counterexamples to Habermas's hypothesis do occur (a case of the second is perhaps provided by the technology of the production of energy from solar and fusion sources, which advanced capitalism seems biased against developing). Moreover, the hypothesis does not square with the possibility that growth in the productive forces could increase the technical feasibility of revolution and thereby increase instability.

As noted earlier, Habermas distinguishes between production and communication. He uses these two concepts to distinguish two ways in which human action can be made more rational (Habermas, 1976, pp. 30-35; 1979, pp. 116-120). In production our concern is the technical one of the selection and organization of means to achieve given ends. Rationalization of this aspect of our lives then consists in making this selection and organization process more effective —in other words, increasing our control over nature. In communication on the other hand, our concern is with achieving, observing and maintaining "action orienting mutual understanding". Rationalization here consists in improving this process, i.e., in internalizing better and better conceptions of ourselves and each other. Habermas calls this concern "practical" rather than technical, because its aim is not the mastery of an objectified process but rather the setting free of "intersubjectivity". Practical rationalization is measured against whether our mutual understanding is based on authentic (sincere, truthful) self-expression and normative behavioral expectations which would be endorsed in unconstrained discussion — i.e., discussion guided solely by the force of the better argument.

In Habermas's theory the terms "technical rationalization" and "practical rationalization" are ambiguous, each having a primary and a secondary sense (Habermas, 1976, p. 235). In the primary sense, technical rationalization takes place when a more efficient technique is placed at our disposal, and practical rationalization when we internalize a better mutual understanding. The secondary sense has to do with social mechanisms for the acquisition and application of technically exploitable knowledge on the one hand, and practical knowledge, on the other. Because he sees the institutions of science and technology in advanced capitalism as optimal mechanisms of the former sort, Habermas views this society as having completed technical rationalization (in the secondary sense). On the other hand, practical rationalization is not yet complete. Institutions do not yet exist which guarantee that practical issues will be resolved discursively — i.e., on the basis of unconstrained discussion by all who are affected by the issue. Thus "the institutionalization of general practical discourse would introduce a new stage of learning for society" (Habermas, 1973a, pp. 29-30; 1975, p. 16).

In this paper we have included both hindering the use and hindering the development of the productive forces as ways in which they can be fettered. We have not yet faced the question, however, of whether "development" and "use" are to be construed purely quantitatively or whether qualitative considerations are also to be allowed. Habermas opts for the former construal, because he views fettering as structural hindrance of technical rationalization in the primary sense, or alternatively incomplete technical rationalization in the secondary sense. (It is only in the secondary sense that one can speak of rationalization as complete, since the possibility remains open that more efficient techniques or better mutual understandings will be found). Productive forces are resources for solving technical problems. The adequacy of a solution to a technical problem is a matter of effectiveness of the solution to achieve the ends which define the problem. Thus society's direction of productive resource use to the wrong (ultimate) ends does not count as fettering of the productive forces. What we have here, Habermas would say, is a *misapplication of*, rather than an *inadequacy in*, society's technical reason. Hindrance of productive resource use counts as fettering only when it prevents use which is optimal with respect to ends actually pursued. Similar remarks apply to the concept of development of the productive forces. It is of course true, Habermas claims, that in advanced capitalism what capacities are developed is not an issue which is resolved through democratic decision-making processes, but this is a matter of incomplete practical rationalization, not incomplete technical rationalization.

In the following discussion, Habermas's view of fettering the productive forces as essentially a matter of incomplete technical rationalization will not be questioned. What will be argued rather is that the completion of technical rationalization is incompatible with capitalist (even *advanced* capitalist) relations of production. Three arguments will be presented. The first has a dialectical character: it points to certain elements of Habermas own theory of truth which speaks against the reconcileability of his claim that technical rationalization is complete with his claim that practical rationalization is *incomplete*. The remaining two criticize his notion of technical rationalization and defend the claim that a more accurate account brings to light certain barriers which capitalism places against technical rationalization.

By Habermas's own account (e.g.,1973b) knowledge of natural truth, the sort of knowledge which is technically utilizable, is optimally acquired when hypotheses are evaluated discursively, i.e., on the basis of discussion guided solely by the force of the better argument. Since such discussion is concerned primarily with establishing the truth of propositions, Habermas calls it "theoretical" discourse, to distinguish it from discursive determination of the correctness of norms, which he calls "practical" discourse. Yet, in spite of the fact that he speaks of natural and moral claims as having different "logics of justification", he asserts both that discursive evaluation of a scientific hypothesis may require consideration of normative issues (e.g., what ends ought to guide knowledge) and that discursive evaluation of a rightness claim may require consideration of natural truth claims (concerning, e.g., what we can know). If at these levels the boundaries between theoretical and practical discourse break down, how can one type of discourse be institutionalized while the other is not? If, as McCarthy puts it, "theoretical and practical reason are inextricably linked" (1978, p. 317), how can one be realized without the other?

We have interpreted Habermas as measuring technical rationalization against the achievement of goals actually pursued and as placing all criticism of those goals in the domain of the practical. However, there is a criticism of goals actually pursued which clearly seems technical character: namely, that those goals do not coincide with the goals which, if achieved, would accord the appropriate satisfaction. This criticism does not rest on a rejection of the view that the adequacy of a solution to a technical problem is a matter of effectiveness of the solution to achieve the ends which define the problem; rather it construes those ends as the goals which would actually accord satisfaction rather than the goals which, because of the lack of the appropriate factual (non-normative) information, the agent in question wrongly thinks would provide satisfaction. Perhaps Habermas would agree with this construal; if so, then our earlier interpretation is incorrect. But this construal makes it much harder to justify the claim that advanced capitalism does not fetter the productive forces. A good case can be made that advertising and other aspects of modern capitalist culture augment the discrepancy between the two sorts of goals distinguished above. Partly on the basis of such an argument, Cohen (1978) constructs a case for the claim that "the structure of the economy militates against optimal use of its productive capacity" (p. 310) in that it biases the application of productive-capacity increases toward the expansion of output rather than toward the now more desirable reduction of toil.

Technical rationality has what might be called a strategic component, which comes into play in situations in which the outcome of one's action depends not only on what one does but on what others do as well. Strategic rationality requires taking this interdependence into account with the aim of best realizing one's ends. Habermas's account of strategic rationality is deficient. Two of his remarks in particular are important. The first is his stipulation that in strategic action, "each subject is following preferences and decision maxims that he has determined for himself — that is, monologically, regardless of whether or not he agrees therein with other subjects". Second, he asserts that as opposed to communicative action, in strategic action "the truthfulness of expressed intentions is not expected" (1976, p. 33; 1979, pp. 118-9). These remarks arbitrarily preclude behavior that in certain strategic situations would lead to a better realization of each agent's ends. Furthermore, the strategic (and hence technical) rationality which this behavior exhibits is not completely realizable under capitalism.

These points are perhaps best argued by reference to two game-theoretic situations: the Prisoners' Dilemma (PD) and the Assurance Game (AG).[8] In each situation there are two players, A and B, each of whom has to choose exactly one of two strategies, 1 and 2. Let a_ib_j stand for A following strategy i and B following strategy j. In PD the preference orderings of A and B over the possible outcomes are respectively (in descending order):

(PD)
$$A: \quad a_2b_1, \; a_1b_1, \; a_2b_2, \; a_1b_2$$
$$B: \quad a_1b_2, \; a_1b_1, \; a_2b_2, \; a_2b_1$$

In AG the rankings are the same except that the first two entries are switched:

(AG)
$$A: \quad a_1b_1, \; a_2b_1, \; a_2b_2, \; a_1b_2$$
$$B: \quad a_1b_1, \; a_1b_2, \; a_2b_2, \; a_2b_1$$

In non-cooperative games, each player chooses his or her strategy in isolation, without any "preplay communiation" with the other players (Luce and Raiffa, 1957, p.89). No type of collusion, such as joint choice of strategy, binding contracts and side payments, takes place (Owen, 1968, pp. 137, 140). In each of the above situations, the outcome of the non-cooperative game can be suboptimal.

In PD, strategy 2 is a dominant strategy for each player, i.e., it is best for each player regardless of what the other does. Where each player must decide on a strategy in isolation, the outcome will thus be a_2b_2, which is suboptimal, since both A and B prefer a_1b_1.[9] An enforced agreement by A and B each to follow strategy 1 would make each better off, but such a collective choice of strategy is precluded by Habermas's requirement that decisions be made "monologically".

In AG, each person would prefer to follow strategy 1 if he or she were assured that the other would do so as well, but would prefer not to take that course of action if he or she felt the other would not either. Each would prefer to do his or her part rather than be a free rider but would prefer not to be the only one to contribute.

In contrast to PD, in AG the outcome of the non-cooperative game does not have to be suboptimal. Whether it is depends on what each player expects about the other's action (Sen, 1967, p. 122). For example, if the preferences of each player are "common knowledge" (see Lewis, 1969, pp. 52-60) in the sense that everyone knows them, everyone knows that everyone knows, and so on, then presumably each player will follow strategy, even when deciding in isolation, since each expects the other to do so and no one has anything to gain from doing otherwise. The outcome will thus be a_1b_1, which is highest in each player's ranking.

Departures from this condition of common knowledge, however, can produce suboptimalities in the non-cooperative game. An example is provided by Elster (1978, p. 154):

If, in the Assurance Game, each player knows his own payoffs only, but suspects that the payoff structure of the other player is that of the Prisoner's Dilemma, or that the other player suspects his structure to be that of the Prisoner's Dilemma, each will expect the other to choose his second strategy and will therefore have to choose his second strategy himself.

The result in this case is the suboptimal a_2b_2.

A communication structure such as Habermas's "ideal speech situation" (1973b, pp. 252-260), in which authentic self-representation is guaranteed, would allow each player to opt for strategy 1, confident in the expectation that the other will do so as well. This solution, however, violates Habermas's assumption that "truthfulness of expressed intentions is not expected". It also exceeds the level of practical rationalization which according to Habermas can be achieved within capitalism. Of course, merely citing possible solution incompatible with capitalism does not show that capitalism inhibits technical rationalization; there may be other possible solutions compatible with capitalism. But that Habermas's characterization of strategic action rules out a solution which is incompatible with capitalism does indicate that his conception of technical reason has an ideological cast.

In fact, there is another possible solution, but it is not one which will help Habermas's case. It requires transforming the non-cooperative context into a cooperative one. The optimal outcome a_1b_1 could be reached through mutual agreement to follow strategy 1.[10] Thus, in both PD and AG, improvements over the outcome of the non-cooperative game can be brought about by "preplay communication". But Habermas's assumption that strategic decisions are made monologically precludes such solutions and effectively restricts strategic rationality to the precepts of the theory of non-cooperative games. In non-cooperative games, allocation of resources proceeds via privately made decisions. This is the sort of allocation characteristic of commodity production, but it is not a universal or inevitable form. Habermas's restriction thus again steers his conception of technical rationalization in an ideological direction: a particular form of maximizing the realization of given ends — one based on privatistic allocation — is presented as universal. It is not surprising then that he does not see modern capitalism as fettering the productive forces.

An objection along the following lines is likely to arise at this point: Let us concede that Habermas's account of strategic action is deficient in the above described ways. It does not follow that correcting the account requires conceding that advanced capitalism places limits on the growth of technical rationality. For it is a crucial feature of *advanced*, as opposed to *liberal* capitalism that the state *intervenes* in the economy; that is, in some cases it departs from privatistic allocation. There is no reason why the advanced capitalist state cannot impose collective solutions to the above problems so as to eliminate the suboptimalities.

But there *is* such a reason: in many cases, such state-imposed solutions would conflict with the requirements of capital accumulation, the *sine qua non* of a healthy capitalist economy. I shall try to illustrate this point with an example discussed by Schweickart (1977, pp. 10-12). It has to do with consumption externalities — effects of an individual's consumption on the well-being of other individuals. Such externalities, for example, exist in the consumption of beverages in disposable bottles: a rise in litter, a decline in the availability of natural resources, and a rise in the cost of municipal waste disposal are all results of such consumption which affect non-consumers as well as consumers. Each person might well choose to buy beer in disposable bottles when confronted with the option as an individual consumer, on the grounds that the convenience is worth the small harmful effect on the environment of such an action. It might also be true that each person would prefer such bottles not to be used at all to everyone's using them. Nevertheless, the latter state of affairs will be the result of the privately made decisions of each consumer. If each individual prefers the state of affairs in which he or she is a "free rider", i.e., the only individual consuming beer in non-returnables, to the state of affairs in which no one is doing so, then this example is a case of the Prisoners' Dilemma (suitably generalized to allow for more than two players); if everyone has the reverse preferences, then we have an Assurance Game (suitably generalized). Market allocation can be suboptimal in these cases. In the presence of consumption externalities, "it might be the case that nobody, not even a tiny minority, preferred the production of X to its ban, and yet the market would supply X to everyone".

The solution to this problem might seem to be at hand in advanced capitalism: a legislative prohibition of the production of beer in non-returnable bottles. However, let us suppose that in terms of private costs only — i.e., monetary costs to individual capitalist enterprises — it is cheaper to produce beer in nonreturnables than in returnables. Under these circumstances, the proposed legislation would reduce the potential for capital accumulation and thereby possibly endanger the functioning of the economy. Thus when the social cost represented by inefficiency can be reduced only by increasing private costs of production, it is not at all clear that the state will opt for eliminating the inefficiency; for these circumstances may well impose a choice between social optimality and a "healthy" economy. It is arguable that precisely this dilemma is at the heart of struggles over environmental-protection legislation today. It is a dilemma which is imposed by the capitalist economic structure, with its distinction between private and social costs and its inherent need for continual accumulation.

IV

The principal claims defended in this paper are the following: In building his case for a legitimation crisis, Habermas relies on the premise that widespread political participation would result in a withholding of legitimation for the state's policies aimed at securing capital accumulation, *even if* these policies would be effective. This premise presupposes that the state is no longer in the classical sense superstructural with respect to capital. Habermas's main argument for this presupposition wrongly treats it as following simply from the political character of the relations of production in modern capitalism. Habermas also seems to overestimate the ability of the formally democratic political institutions characteristic of advanced capitalism to sustain discursive evaluation of state economic policy; a proper appraisal of ideological and economic barriers to such evaluation might make his premise appear less plausible. Moreover, Habermas fails to make a good case for the claim which precipitated the search for new forms of crisis-susceptibility: that advanced capitalism does not fetter the productive forces. His negative arguments for this claim, which are aimed at rejecting Marx's crisis theory as inapplicable to advanced capitalism, do not work: that the state is not superstructural does not entail that Marx's theory is inapplicable, but only that it is possibly insufficient to account for all crisis tendencies; and the systematic production in advanced capitalism of productivity-increasing innovations cannot vitiate Marx's theory of the tendency of the rate of profit to fall, if that theory was true of liberal capitalism, since the theory asserts a crisis-inevitability for capital *per se* and not just for some particular variant for it. Moreover, the support for the claim that advanced capitalism does not fetter the productive forces which Habermas derives from his theory of social evolution is spurious, since it rests in part on the questionable hypothesis that development of the productive forces always works *(ceteris paribus)* to increase social stability and in part on an inadequate account of the nature of technical rationality. The requirements of capital accumulation can conflict with the demands of technical reason as well as the demands of practical reason.

Despite the largely negative thrust of these results, Habermas's research program provides important input into the contemporary Marxist discussion of the nature of advanced capitalism and its place in history. The significance of his work does not derive from his having come to grips with a mutation in capitalism which solves the problem of economic crises, for there has been no such mutation. Rather, Habermas speaks in an informed, systematic, and provocative way to other important issues.

An unsettled problem in Marxian theory concerns the status which should be accorded moral claims. Does Marx's critique of capitalism appeal to moral standards, or are purely non-moral norms at the bottom of it? Whatever these norms are, how are they to be justified? Habermas's work offers us a well-articulated, plausible position on these issues; critical discussion of it cannot but help to resolve them[11].

Habermas's distinction between technical and practical rationalization forces us to pay attention to the fact that human beings develop not only capacities to change external nature but also capacities to change themselves — their self-conceptions and the norms they live by. Whether these are fundamentally distinct kinds of capacities, having, as Habermas puts it, distinct logics of development, or whether, as Marx seemed to think, their development can be subsumed under a single concept of rationalization, is very much an open and important question. Even if the single-logic view proves to be correct, the nature of this logic remains to be worked out, and useful input is likely to come from Habermas's work.

Another element of Marxism crying out for development is the theory of revolutionary motivation. Buchanan (1979) has pointed to what are apparently Prisoners'-Dilemma and perhaps Assurance-Game situations faced by members of the working class in deciding whether to engage in revolutionary activity. In the last section I suggested that what Habermas calls practical rationalization provides a way of dealing with such problems. Perhaps Habermas's work contains the seeds of an adequate theory of revolutionary motivation. This is obviously one of Habermas's own concerns.

Historical materialism, though it has proved to be an enormously fruitful heuristic device, is still very much an embryonic theory. Habermas correctly stresses the crucial importance of developing it further. We have seen that it has important implications for the crisis susceptibility of modern capitalism. Habermas also mentions the difficult problem of individuating social formations as one which a fuller working out of historical materialism's concepts and claims is likely to help solve. I would like to close by mentioning one further problem. Marxism has yet to develop an adequate explanation of one crucially important feature of capitalism: its patriarchal character. It seems clear from recent developments in feminist theory that such an explanation will have to refer not only to the capitalist production of commodities but also to the practices of bearing and rearing children and the provision of affection, nurturance and sexual satisfaction[12]. A materialist account of male domination requires an understanding of the nature of these practices, in particular of the social mechanisms governing (1) the acquisition, use and reproduction of the abilities and resources utilized in them and (2) the distribution of their fruits. Habermas's efforts to work out a typology of human action can almost certainly throw some light on these questions. There is in fact an interesting parallel between Habermas and earlier Frankfurt-School theorists on the one hand, and materialistically inclined feminist theorists on the other. Both have claimed that what classical Marxism has focused on as production is too narrow a basis for understanding and appraising the development of social formations through history. This claim may provide the basis for a fruitful dialogue.

NOTES

1. "Advanced capitalism is certainly not characterized by a fettering of the productive forces" (Habermas, 1976, p. 53). See also Habermas (1968, p. 99; 1970, p. 119).

2. By the "classical version" I mean the view which "became the guiding principle" of Marx's studies (Marx, 1859, p. 20). A great deal of progress in the articulation of this view has recently been made by Cohen (1978) and McMurtry (1978). The concept of the superstructural in what follows largely reflects their influence.

3. See Cohen (1978, ch. VIII) for the importance of distinguishing rights and powers and for an explanation of the frequent use of legal language in specifying economic structure.

4. In Sensat (1979, ch. 6), I try to demonstrate and account for a deficient appreciation in Habermas's work of these mystifications.

5. For attempts to locate these crisis periods historically and to describe the social restructuring which led to their resolution, see, for example, Hobsbawm (1975) and Wright (1978, ch. 3).

6. See, for example, Laibman (1976, 1977), Roemer (1979), Shaikh (1978), Steedman (1980), Nakatani (1980), Armstrong and Glyn (1980), Bleaney (1980) and Shaikh (1980).

7. For an elaboration of the claims made in this paragraph, see Sensat (1979, ch. 7). Wright (1978, ch. 3) suggests that the various crisis-inducement mechanisms (e.g., rising organic composition, rising labor costs, realization failure) stressed in the different Marxist theories of economic crisis should be seen as themselves reflections of underlying contradictions of capital and that which of these can be expected to come into play depends on peculiarities of the social structure of accumulation in existence. For an empirical study of the role of these mechanisms in the postwar U.S. economy, see Weisskopf (1979).

8. The Prisoners' Dilemma is due to A. W. Tucker (Luce and Raiffa, 1957, sec. 5.4). The Assurance Game seems to have been introduced into the literature by Sen (1967, 1969). The orderings of possible outcomes in the version of the Assurance Game in Sen (1969) differs slightly — but immaterially for present purposes — from the version presented here, which follows Sen (1974).

9. We ignore possible exceptions to this suboptimality associated with temporal repetition of the game. See Luce and Raiffa (1957, sec. 5.5) and Taylor (1976).

10. Where deception or self-deception on the part of the players is suspected as a possibility, enforcement of the agreement might be necessary.

11. Important papers in this area are those by Wood (1972, 1979), Husami (1978) and Nielson (1980).

12. Rubin (1975), Dinnerstein (1977), Chodorow (1978), Hartmann (1979) and Ferguson and Folbre (1979).

REFERENCES

1. Armstrong, P. and Glyn, A. (1980), "The Law of the Falling Rate of Profit and Oligopoly: A Comment on Shaikh", *Cambridge Journal of Economics*, vol. 4, pp. 69-70

2. Bleaney, M. (1980), "Maurice Dobb's Theory of Crisis: A Comment", *Cambridge Journal of Economics* vol. 4, pp. 71-3

3. Buchanan, A. (1979), "Revolutionary Motivation and Rationality", *Philosophy and Public Affairs* vol. 9, pp. 59-80

4. Chodorow, N. (1978), *The Reproduction of Mothering*, Berkeley, Los Angeles and London: University of California Press

5. Cohen, G.A. (1978), *Karl Marx's Theory of History: A Defence*, Princeton: Princeton University Press

6. Dinnerstein, D. (1977), *The Mermaid and the Minotaur*, New York: Harper Colophon Books

7. Edwards, R.C. and Reich, M. (1978), "Party Politics and Class Conflict", in: *The Capitalist System* (ed. R.C. Edwards, M. Reich and T.E. Weisskopf), Englewood Cliffs: Prentice Hall

8. Elster, J. (1978), *Logic and Society* Chichester, New York, Brisbane, Toronto: John Wiley and Sons

9. Ferguson, A. and Folbre, N. (1979), "The Unhappy Marriage of Patriarchy and Capitalism", Mimeo

10. Gordon, D.M. (1980), "Stages of Accumulation and Long Economic Cycles", in: *Processes of the World System* (ed. T. Hopkins and I. Wallerstein), Beverly Hills and London: Sage Publications

11. Habermas, J. (1968), *Technik und Wissenschaft als "Ideologie"*, Frankfurt: Suhrkamp

12. Habermas, J. (1970), *Toward a Rational Society* (tr. J. Shapiro), Boston: Beacon Press

13. Habermas, J. (1973a), *Legitimationsprobleme im Spätkapitalismus*, Frankfurt: Suhrkamp

14. Habermas, J. (1973b), "Wahrheitstheorien", in: *Wirklichkeit und Reflexion: Walter Schulz zum 60. Geburstag* (ed. Helmut Fahrenbach), Pfullingen: Gunther Neske

15. Habermas, J. (1975), *Legitimation Crisis* (tr. T. McCarthy), Boston: Beacon Press

16. Habermas, J. (1976), *Zur Rekonstruktion des Historischen Materialismus*, Frankfurt: Suhrkamp

17. Habermas, J. (1979), *Communication and the Evolution of Society* (tr. T. McCarthy), Boston: Beacon Press

18. Hartmann, H. (1979), "The Unhappy Marriage of Marxism and Feminism: Towards a More Progressive Union", *Capital and Class*, no. 8, pp. 1-33

19. Hobsbawm, E. (1975), "The Crisis of Capitalism in Historical Perspective", *Marxism Today*, vol. 19, no. 10

20. Husami, Z. (1978), "Marx on Distributive Justice", *Philosophy and Public Affairs*, vol. 8, pp. 27-64

21. Laibman, D. (1976), "The Marxian Labor-Saving Bias: A Formalization",

Quarterly Review of Economics and Business, vol. 16, pp. 25-44

22. Laibman, D. (1977), "Toward a Marxian Model of Economic Growth", *American Economic Review*, vol. 67, pp. 387-92

23. Lewis, D. (1969), *Convention: A Philosophical Study*, Cambridge, Mass.: Harvard University Press

24. Luce, R. and Raiffa, H. (1957), *Games and Decisions*, New York: John Wiley and Sons

25. McCarthy, T. (1978), *The Critical Theory of Jürgen Habermas*, Cambridge, Mass. and London: MIT Press

26. McMurtry, J. (1978), *The Structure of Marx's World View*, Princeton: Princeton University Press

27. Marx, K. (1857-8), *Grundrisse* (tr. M. Nicolaus), Baltimore: Penguin, 1973

28. Marx, K. (1859), *A Contribution to the Critique of Political Economy* (ed. M. Dobb, tr. S.W. Ryazanskaya), Moscow: Progress Publishers, 1970

29. Marx, K. (1862-3), *Theories of Surplus Value*, part III, Moscow: Progress Publishers, 1971

30. Marx, K. (1894), *Capital*, vol. III, New York, International Publishers, 1967

31. Nakatani, T. (1980), "The Law of Falling Rate of Profit and the Competitive Battle: Comment on Shaikh", *Cambridge Journal of Economics*, vol. 4, pp. 65-8

32. Nielson, K. (1980), "Marxism, Ideology, and Moral Philosophy", *Social Theory and Practice*, vol. 6, pp. 53-68

33. Owen, G. (1968), *Game Theory*, Philadelphia, London, Toronto: W.B. Saunders Company

34. Rubin, G. (1975), "The Traffic in Women: Notes on the 'Political Economy' of Sex", in: *Toward an Anthropology of Women* (ed. R. Reiter), New York: Monthly Review Press

35. Schweickart, D. (1977), "Worker-Controlled Socialism: A Blueprint and a Defense", *Radical Philosophers' Newsjournal*, vol. 8, pp. 1-22

36. Sen, A. (1967), "Isolation, Assurance and the Social Rate of Discount", *Quarterly Journal of Economics*, vo, 81, pp. 112-124

37. Sen, A. (1969), "A Game-Theoretic Analysis of Theories of Collectivism in Allocation", in: *Growth and Choice* (ed. T. Majumdar), London: Oxford University Press

38. Sen, A. (1974), "Choice, Orderings and Morality", in: *Practical Reason* (ed. S. Körner), Oxford: Basil Blackwell

39. Sensat, J. (1979), *Habermas and Marxism: An Appraisal*, Beverly Hills and London: Sage Publications

40. Shaikh, A. (1978), "Political Economy and Capitalism: Notes on Dobb's Theory of Crisis", *Cambridge Journal of Economics*, vol. 2

41. Shaikh, A. (1980), "Marxian Competition versus Perfect Competition: Further Comments on the So-Called Choice of Technique", *Cambridge Journal of Economics*, vol. 4, pp. 75-83

42. Steedman, I. (1980), "A Note on the 'Choice of Technique' under Capitalism", *Cambridge Journal of Economics*, vol. 4, pp. 61-4

43. Taylor, M. (1976), *Anarchy and Cooperation*, London: Wiley

44. Weisskopf, T. (1979), "Marxian Crisis Theory and the Rate of Profit in the Postwar U.S. Economy", *Cambridge Journal of Economics* vol. 3, pp. 341-78

45. Wood, A. (1972), "The Marxian Critique of Justice", *Philosophy and Public Affairs*, vol. 1, pp. 244-82

46. Wood, A. (1979), "Marx on Right and Justice: A Reply to Husami", *Philosophy and Public Affairs*, vol. 8, pp. 267-95

47. Wright, E.O. (1978), *Class, Crisis and the State*, London: NLB

DISCUSSION

Jerzy Brzeziński/Poznań

A STATISTICAL MODEL
OF DATA ANALYSIS
IN INTERACTIONAL PSYCHOLOGY
Comments on the quantitative analysis of the scores
of the "S-R" Inventory of Anxiousness

1. Introduction

In the last thirty years many inventories for measuring the variable of
anxiety have been developed. The most popular ones are: Taylor (1953)
Manifest Anxiety Scale (MAS), Spielberger, Gorsuch, and Lushene
(1970) State-Trait Anxiety Inventory (STAI), Cattell and Scheier (1967)
IPAT Anxiety Scale, as well as inventories based upon the theoretical
assumptions of interactional psychology like Endler, Hunt, and
Rosenstein (1962) S-R Inventory of Anxiousness; Endler and Okada
(1975) S-R Inventory of General Trait Anxiousness.

It would seem to be of particular advantage for scientific and
diagnostic practice to connect the theoretical and methodological
conception of so-called interactional psychology, worked out by
Magnusson and his co-workers, with the theory of anxiety (and its
measurement) worked out by Spielberger and his colleagues, namely the
conception of "state and trait anxiety" (Spielberger, 1975a, 1975b). The
relationship between the S-R Inventory of General Trait Anxiousness,
based on the theoretical assumptions of interactional psychology, and
Form A-Trait STAI Spielberger et al. has already been pointed out by
Endler and Okada (1975).

In the present paper I would like to pay attention to the
methodological and statistical methods of measuring anxiety in
inventories of the S-R type. The methodological and statistical solutions
proposed by Endler et al. (1962) cannot be fully accepted. Scientific and
diagnostic practice should be built upon the theoretical and
methodological conceptions to be as effective as possible, we have to
eliminate from the "basis" these assumptions which are formulated in a
wrong way. And while I can fully accept the theoretical bases of
interactional psychology, some of the statistical assumptions (accepting
them influences the results of the statistical analysis of data obtained with

the help of the inventory of anxiety) need to be corrected. These corrections will contribute to the increase of the objectivity of data obtained with the help of the inventories of anxiety based upon the assumptions of interactional psychology.

The psychological conception called "interactional psychology", developed by Magnusson, Endler and others, was meant to solve the problem of factors that determine human behavior. In opposition to the classical approaches, stressing either the superiority of situational factors (*situationism*, according to Ekehammar, 1974), or domination of subjective factors in explaining human behavior (*personologism*, according to Ekehammar, 1974), Magnusson and others have formulated the conception of *interactional psychology*. Obviously enough, their purpose was to overcome certain theoretical difficulties engendered by the conceptions mentioned above, which have approached the determinants of human behavior very one-sidedly.

The concept of interaction is explicated in psychology in many different ways (Olweus, 1977). Main theoreticians of that direction, Magnusson and Endler (1977), concentrated the possible interpretations of the basic concepts of interactional psychology around two fundamental meanings. According to the first one, the concept of "dynamic interaction" is introduced in the sense of mutual relations occurring between situational and subjective factors. These relations are of a procedural character, since we are dealing here with the influences of a situation, occurring in time, upon a person, as well as the reverse. According to the other one, which may be called "statistical", interaction is viewed in the sense of the ANOVA model.

The statistical view of interaction has somehow determined the approach to planning empirical investigations, the construction of techniques of accumulating data and the testing of hypotheses. Leaving the technical details concerning the ANOVA model aside, I would only like to point the readers's attention to the fact that, according to its internal logic, explaining the total variance of the dependent variable Y is carried out by means of disarraying the total variance of variable Y and splitting it into variance components. The source of those components are particular independent variables (or their classes) and their interactions. An additional source of a component of variance is the influence of interfering variables, uncontrolled by the researcher. The effect of their activity is expressed in the *residual variance*.

According to Magnusson and Endler, there are three basic sources of variance components (apart from the residual variance), which the investigator is trying to discover. These are: (a) a situational differentiation which manifests itself through the abundance of real life situations in which man functions — *the situational variable* (b); individual differences (or: interpersonal differentiation) — *the subjective variable*; (c) *the interaction of (a) and (b)*.

In order to answer the question of what the split of the total variation Y into variance components looks like, Endler, Hunt, and Rosenstein (1962) prepared a prototype of a new inventory (of an "S-R" type), which has become a model solution and is used in different forms by the investigators of Magnusson and Endler's school. The peculiarity of that solution is the fact that its arrangement has been subordinated to the *ANOVA three-factor design* of $n = 1$ (*random effects model*, or *II*). The results of testing a specified random sample representative of a given population, using that inventory, are, in turn, exposed to a statistical analysis. This analysis considers as a starting point the distribution of the values of *expected mean squares*, the so-called $E(MS)$, within the framework of the assumed ANOVA design. It allows for a proportional specification of the values of particular variance components related to the total variation of variable Y (assumed as 100%). The algorythm of the procedure has been borrowed from Gleser, Cronbach and Rajaratnam (1965), together with the later developments of Endler (1966).

Before I carry out an analysis of the methodological and statistical solution adopted by the investigators representing the above-mentioned theoretical orientation, I shall give a brief presentation of the core of the prototype version of the "S-R" type questionnaire, worked out by Endler et al. (1962). Despite the fact that this publication appeared over 20 years ago, the formal arrangement of the questionnaire introduced then is still followed by many interactionists (e.g., Endler and Okada, 1975).

The original version of the "S-R" Inventory of Anxiousness includes two lists:

(a)   a list of 11 situations containing different types of threat
(b)   a list of 14 ways of reacting to these.

Below is a sample of two situations from (a):

1.   You are just starting off on a long automobile trip.
3.   You are going into a psychological experiment.

The inventory has the form of an 11-page work book. Each page contains all 14 possible reactions together with a 5-point rating scale for each. The given situation is presented at the top of the page. Below are a few lines from the first page providing a sample of several reactions from (b):

"You are just starting off on a long automobile trip.

1.   Hart beats faster: not at all 1 2 3 4 5 much faster
5.   Want to avoid situation: not at all 1 2 3 4 5 very much."

The subject's task is to mark on each rating scale the intensity of a given reaction (one of the 14) in relation to each of the 11 situation, separately. After each of the subjects, selected from a sample representative of a given population, has filled out the work book, we shall have obtained from him a matrix of results of the size 11 x 14. All outcomes thus obtained (their number is 11 x 14 x *the number of persons* (*n*)) are subjected to analysis of variance.

Endler et al. (1962) have distinguished the following sources of variance:

1. factor A — *Subjects* — *Ss.*
2. factor B — *Situations* — *Sits.*
3. factor C — *Modes of response* — *M-R.*
4. *Ss x Sits.*
5. *Ss x M-R.*
6. *Sits x M-R.*
7. *Residual factors*, i.e., factors other than the above ones, also including the interaction A x B x C.

Corresponding to the above-mentioned sources of variance is a three-factor ANOVA design (according to model *II*) with $n = 1$ per cell.

The use of ANOVA in questionnaire testing is inevitably connected with the problem of finding appropriate indicators for the evaluation of the value of dependent variable variance, connected in turn, with its controlled and carried out by the investigator according to a specified ANOVA design. Out of several different indicators of variable significance, many authors (cf. Gaito, 1960; Scheffé, 1959; Fleiss, 1969; Cronbach, Glesner, Nanda and Rajaratnam, 1972, Endler, 1966; Vaughan and Corballis, 1969) have suggested in recent years that the total variation be split into components and these, in turn, should be evaluated by means of an analysis of the values of the expected mean squares, $E(MS)$.

In the work of Vaughan and Corballis (1969), for example, we may find the specifications of $E(MS)$ for three fundamental ANOVA designs — the one-, two-, and three-factor designs (in all three models: *I, II, III*). Also given are the calculation formulae which make it possible to estimate the proportional values of variance components. I shall mention here only these formulae which concern the ANOVA design used by Endler et al. (1962).

Let us adopt the following denotations:

$A = (a_1,..., a_i,...a_p)$ — factor A of p-number of levels
$B = (b_1,..., b_j,...b_q)$ — factor B of a q-number of levels
$C = (c_1,..., c_z,...c_r)$ — factor C of an r-number of levels
A, B, C — "fixed" factors (in the sense of the fixed effects model — *I*).
A', B', C' — "random" factors (in the sense of the random effects model — *II*).

Due to the above, the arrangement of factors A', B, C should be considered within the framework of the mixed effects model — *III* while the arrangement of factors A, B, C, will be considered within the framework of model II.

With the use of the method developed by Cornfield and Tukey (1956), the obtained specification of the expected mean squares for the three-factor analysis of variance design with $n = 1$ (model II) were similar to those obtained by Endler et al. (1962):

$$E(MS_{A'}) = \delta_R^2 + q\delta_{\alpha\gamma}^2 + r\delta_{\alpha\beta}^2 + qr\delta_\alpha^2$$

$$E(MS_{B'}) = \delta_R^2 + p\delta_{\beta\gamma}^2 + r\delta_{\alpha\beta}^2 + pr\delta_\beta^2$$

$$E(MS_{C'}) = \delta_R^2 + q\delta_{\alpha\gamma}^2 + p\delta_{\beta\gamma}^2 + pq\delta_\gamma^2$$

$$E(MS_{A'B'}) = \delta_R^2 + r\delta_{\alpha\beta}^2$$

$$E(MS_{A'C'}) = \delta_R^2 + q\delta_{\alpha\gamma}^2$$

$$E(MS_{B'C'}) = \delta_R^2 + p\delta_{\beta\gamma}^2$$

$$E(MS_R) = \delta_R^2$$

R denotes here the "residual", which means everything that is included in the experimental error and the interaction $A \times B \times C$. With appropriate transformations of E(MS) (see Endler and Hunt, 1966; Endler, 1966) we shall obtain the following indicators of variance components:

$$\hat{\delta}_\alpha^2 = \frac{MS_A - MS_{AB} - MS_{AC} + MS_R}{qr}$$

$$\hat{\delta}_\beta^2 = \frac{MS_B - MS_{AB} - MS_{BC} + MS_R}{pr}$$

$$\hat{\delta}_\gamma^2 = \frac{MS_C - MS_{AC} - MS_{BC} + MS_R}{pq}$$

$$\hat{\delta}_{\alpha\beta}^2 = \frac{MS_{AB} - MS_R}{r}$$

$$\hat{\delta}_{\alpha\gamma}^2 = \frac{MS_{AC} - MS_R}{q}$$

$$\hat{\delta}_{\beta\gamma}^2 = \frac{MS_{BC} - MS_R}{p}$$

$$\hat{\delta}_R^2 = MS_R$$

As a final remark on the "S-R" inventory, I would like to stress that the construction proposed by those investigators is very heuristic and that inventory makes possible investigations into new fields, which is possible, among other things, because of combining them with analysis of variance so commonly used by psychologists.

The "S-R" inventory design is worth popularization as far as constructing new personality questionnaires and constructing a set of quasi-psychometric methods are concerned (it may be regarded as a link between the psychometric and clinical approach in personality research).

The methodology of empirical investigations proposed by Endler et al. (1962) cannot be adopted without any objections, despite its clarity, neat construction, and relationship to the ANOVA model commonly used by psychologists. The remarks that come to one's mind because of the above reason can be presented in two main points:

1. The problem of selecting an ANOVA design adequate to the data structure.

2. The problem of an alternative statistical model (in relation to the ANOVA model) for the analysis of data obtained with the use of a device like the "S-R" inventory.

They will be discussed in two subsequent sections of this paper.

2. The "S-R" inventory and the ANOVA model

As has already been mentioned, in order to statistically work out the data obtained through the use of the "S-R" inventory Endler, Hunt and Rosenstein have used the design of a three-factor ANOVA, with $n = 1$ (model II). This design requires that the subjects be assigned randomly (the randomization principle) to each of the $p\ q\ r$ combinations of the levels of factors A, B, C.[1] If we leave factor A (persons) aside for the time being, the minimum size of the group which should be tested with the questionnaire should amount to 154 persons. However, in fact, the introduction of the third factor, "the subjects" (A), violates the condition of "one cell = one subject". Endler represents a peculiar principle, namely, "one subject = 154 cells", which equals the product of $q\ r$.

What we are dealing with here is the classical case of the ANOVA design with repeated measurements (such designs are discussed by Winer, 1971 and Ferguson, 1976).

Let us, however, in order to continue the discussion, ignore this observation and agree with Endler that his choice of the ANOVA design is correct. This creates another problem. The use of the three-factor design with $n = 1$ would be impossible (no possibility of estimating MS_{error}), if we did not accept the idealizing assumption concerning the interaction of the second order, namely, that $A \times B \times C$ is assumed to be zero. To be more precise, it is assumed that the residual variance (δ_R^2) eqals the sum of the error variance (δ_e^2) and the variance explained by the triple interaction of factors: A, B, C ($\delta_{\alpha\beta\gamma}^2$).

At this point, we can point out another contradiction among the ontological assumptions of interactional psychology. This contradiction lies between the search for the sources of the variances in human behavior in the interaction among the variables significant to Y and the research practice forced out by the form of the ANOVA design adopted by the authors. On the other hand, Endler says (1966, p. 566) that the second order interaction has psychological meaning and illustrates this with an example. On the other hand, however, he agrees with the procedure of combining this interaction with the random error term, and so we may speak of a joint residual variance. If we look at the results of investigations, carried out by Endler and Hunt on the basis of testing a sample of 67 students from Illinois University with the "S-R" inventory, we shall notice that the "residual" (which means error plus interaction $A \times B \times C$) explains almost 30% of the total variance of Y, while the other interactions ($A \times B$, $B \times C$, $A \times C$) explain about 6% to 17% of the total variance of Y (Endler and Hunt, 1966). This would mean that either the investigation had been conducted improperly (the assumption on the zero interaction of factors A, B, C), or that its error is minor and a significant contribution to the residual variance is made by the second order interaction $A \times B \times C$ (the assumption on the non-zero interaction of factors A, B, C). To be in accordance with the theoretical assumptions of interactional psychology, it is impossible (if three-factors are included in the image of the space of variables significant to Y) to accept three interactions of the first order and neglect the interaction of the second order. It should be particularly the latter that ought to explain the largest percentage of the total variance of variable Y.

If the above idealizing assumption is accepted, then it must be MS_{ABC}, viewed as the denominator of the particular relations of F, that is adopted as the estimate of the residual variance. In order to make sure that this simplification is justified, we can refer to an appropriate test. In this case, it is Tukey's test for nonadditivity.[2] The insignificance of the effect of the three-factor interaction, discovered with the use of that test, allows us to use the technical operation described above. In other words, instead the model of:

$$Y_{ijzk} = \mu + a_i + \beta_j + \gamma_z + a_i\beta_j + \beta_j\gamma_z + a_i\gamma_z + a_i\beta_j\gamma_z + \epsilon_{ijz(k)}$$

we accept the following model:

$$Y_{ijzk} = \mu + a_i + \beta_j + \gamma_z + a_i\beta_j + \beta_j\gamma_z + a_i\gamma_z + \epsilon^*_{ijz(k)}$$

Endler is aware of the fact that testing the hypothesis of a zero second order interaction $A \times B \times C$ is possible. Nevertheless, neither does he use Tukey's test, nor does he encourage it because for large samples as Endler writes: "the computation of the nonadditivity sums of squares becomes a tedious and cumbersome task" (1966, p. 566). The remark which characterizes the computation as "tedious" and "cumbersome" cannot be taken seriously. The reason for that is that the technical aspect of using a given statistical procedure cannot determine its usefulness if the assumptions of the ANOVA model require that it should be used. Apart from that, from the technical point of view, the three-factor analysis of variance calculated in a traditional way is also tedious and cumbersome but who would carry out complex calculations without the use of a computer?

Another question that remains is the problem of justifying Endler's (1966) selection of the model of random effects. According to that model, it is assumed that the levels of each of the three-factors *A, B, C* constitute attempts randomly drawn from the population of the possible levels of these factors. This assumption can be accepted in the case of the subjects factor, because it is, in fact, possible to randomly select a sample which would be representative for a given population. However, it cannot be done in the case of the remaining two factors *B* and *C*. They can be only considered as "fixed" factors. And so, instead of the model of random effects, the model of mixed effects should be used. Endler (1966) considered both models in the context of data obtained through the use of the S-R inventory. For technical reasons, however, he decided to select model *II*, even though it is inadequate. These technical reasons involved the inability to obtain explicit (precise) evaluations of particular variance components. If model *II* had not been selected (as was done by Endler) but rather model *III*, the variance components having their source in factors *B* and *C*, as well as some of the interactions, could not be explicitly defined. Again, the desire to make the statistical solution as neat as possible has overshadowed its adjustment to the investigated reality.

In conclusion, we should say that the critique of the methodological and statistical solution put forward by Endler et al. (1962) involves the following:

(a) the selection of an inappropriate ANOVA design (the three-factor one with $n = 1$);

(b) disregarding the assumptions which constitute the selected ANOVA design (the assumption of the additivity of the effects of particular factors);

(c) the selection of an inappropriate ANOVA model (the random one instead of the mixed one).

Finally, I would like to ask a question concerning the number of factors used by Endler et. al. (1962) in relationship to the dependent variable. The question is whether we are really dealing with three different factors. There is no doubt about the factor "situations", or the factor "modes of response". These have been clearly distinguished in the "S-R" inventory and the number of categories in each factor has also been precisely specified. Also the interaction "situations x modes of response" has psychological meaning. By comparing different populations (students, soldiers, patients, etc.), we may answer questions concerning the relative size (in relation to a given population) of the interaction: "SxR" and its role in explaining the variance in the dependent variable. The introduction of the third factor "subjects", in which the number of categories is equal to the number of persons in the sample somehow obscures this clear picture. It is difficult to interpret it psychologically as a source of variance. The reason for this is that, if the investigations were carried out on a sample representative of a given population (Endler et al. has used the populations of students of certain American universities), then this sample ought to be homogeneous. Homogeneity means here a lack of systematic intra-group differentiation. In other words, together with the increase in the group homogeneity, the intra-group variance decreases. Thus one should not expect a significant participation of that factor in explaining the total variance of variable Y. One way or another, the value of the variance components is here a function of group homogeneity which, in turn, is dependent upon the method of selecting the sample from the population. Therefore, different values for the variance components introduced by the factor "subjects" (obtained by different investigators) can be explained by reference to the technique by which the sample was drawn from the population. And yet, from the point of view of interactional psychology, this conclusion seems trivial.

Selecting only the first two factors leaves the choice of an appropriate ANOVA design still open. The design which I shall present below is not open to the critical remarks presented above. It is a two-factor design with repeated measurements on two factors. The subjects are treated here as a random *quasi-factor*. "Situations" and "Modes of response" are fixed factors. Thus, we are dealing with mixed effects of the ANOVA model (model *III*). This design has been described by Fergusson (1976, p. 309 — 313). Structurally speaking, it corresponds to the design used by Endler et al. (1962). Table 1 is the incorporation of the data obtained through the use of the "S-R" inventory in the design discussed here, while table 2 is the table of the analysis of variance, in the rows of which are particular sources of the variance in the dependent variable.

Table 1

Two-factor design with repeated measures

subjects	Situations – B													
	b_1				b_2 ...					b_{11}				
	Modes of response – C													
	c_1	c_2	c_3	...	c_{14}	c_1	c_2	c_3	...	c_{14}	c_1	c_2	c_3	... c_{14}
1						...								
2						...								
3						...								
i														
p						...								

Table 2

Analysis of variance table for two-factor design with repeated measures (mixed model).

Source of variation	SS	df	MS	F
(1) *Between subjects (A)*				
(2) *Within subjects*				
(3) Situations (B)				
(4) Situations x Subjects				
(5) Modes of response (C)				
(6) Modes of response x Subjects				
(7) Situations x Modes of response (B x C)				
(8) Situations x Modes of response x Subjects				
(9) *Total*				

The distribution of the values of the expected mean squares for that ANOVA design (model *III*) can be presented as follows:

$$E(MS_{A'}) = \delta_\epsilon^2 + qr\delta_a^2$$
$$E(MS_B) = \delta_\epsilon^2 + r\delta_{a\beta}^2 + pr\theta_\beta^2$$
$$E(MS_C) = \delta_\epsilon^2 + \cdot q\delta_{a\gamma}^2 + pq\theta_\gamma^2$$
$$E(MS_{BC}) = \delta_\epsilon^2 + \delta_{a\beta\gamma}^2 + p\theta_{\beta\gamma}^2$$
$$E(MS_{A'B}) = \delta_\epsilon^2 + r\delta_{a\beta}^2$$
$$E(MS_{A'C}) = \delta_\epsilon^2 + q\delta_{a\gamma}^2$$
$$E(MS_{A'BC}) = \delta_\epsilon^2 + \delta_{a\beta\gamma}^2$$

where:

A' — the subjects (the random quasi-factor)
B — situations (the fixed factor)
C — modes of response (the fixed factor)

$$\theta_\beta^2 = \frac{q\delta_\beta^2}{q-1}$$

$$\theta_\gamma^2 = \frac{r\delta_\gamma^2}{r-1}$$

$$\theta_{\beta\gamma}^2 = \frac{qr\delta_{\beta\gamma}^2}{(q-1)(r-1)}$$

Even a first glance at the specification of the above formulae allows to discover that, contrary to Endler's desire[3], we will not obtain explicit evaluations of particular variance components. What we can do is to underestimate a given variance component (or, in other words, to specify its theoretically possible bottom value) at one time, and overestimate it (to specify its theoretically possible top value) at other times. If we are satisfied with such a solution, then, ignoring other critical remarks concerning the solution accepted by Endler, we must consider this ANOVA design as adequate for this type of questionnaire investigations.

3. The S-R inventory and the multiple regression model

It has been settled in the final part of the previous section that the image of the space of variables significant to the dependent variable includes, apart from the two factors (the situational one — (X_1) and the modes of response one — (X_2), also the interaction between these two factors: X_1X_2. My proposal is to replace the ANOVA model with a multiple regression model. It will be a non-additive model (because of the occurence of the interaction between the independent variables) of the following form:

$$Y' = b_{Y_{1.23}}X_1 + b_{Y_{2.13}}X_2 + b_{Y_{3.12}}X_1X_2 + a$$

160

In this model, the correlation between variable X_1 and variable X_2 is called a "moderator variable". In other words, there occurs the following relationship:

$$r_{YX_1} = g(X_2)$$

The non-additive model constitutes an expansion of the additive model:

$$Y' = b_{Y_{1.2}} X_1 + b_{Y_{2.1}} X_2 + a$$

Variables X_1 and X_2 are *qualitative variables* (measured on a nominal scale) and their introduction into the regression equation requires a recoding operation, for example, the system of "orthogonal coding". Also the interaction $X_1 X_2$ requires encoding.

One must also take into account the fact that particular categories of both qualitative variables are not ascribed to different persons, but that each subject "passes" through all the pq categories. And so, as in the case of the ANOVA model, the appropriate design will be one which will permit the repetition of the measurements of the dependent variable within the range of two factors. The statistical bases for this ("repeated") variation of the multiple regression model can be found in Cohen and Cohen (1975).

We can use the coefficient of partial determination (pq^2) or the coefficient of semipartial determination (sr^2) as the measurements of the proportional participation of particular sources of variance in explaining the total variance in variable Y.

I believe that the briefly discussed here variant of the multiple regression model relieves the investigations which use the "S-R" inventory from the problems of a statistical nature mentioned above.

Summary

This article discusses some methodological aspects of the approach of Magnusson, Endler and others to interactional psychology. In particular, it focuses on the use of the ANOVA model in the investigations carried out by Endler. Having presented the content of the investigative procedure within interactional psychology, the author has commented on the methodological and statistical solutions introduced by Endler. According to the author, the use of the ANOVA model for a statistical description of data obtained through the "S-R" inventory (the basic investigative method in interactional psychology) is not appropriate. The following reasons contribute to that conclusion: (a) the selection of an inadequate ANOVA design, (b) making reference to an inadequate ANOVA model. Instead of a three-factor ANOVA design with $n = 1$, a different ANOVA design with repeated measures on the dependent variable within the scope of two factors should rather be used. Moreover, instead of model *II*, ANOVA model *III*s should be employed. All that influences the establishment of the dimensions of particular variance components calculated with the method of the analysis of the expected mean squares values, $E(MS)$. In the final conclusions, the author suggests that the ANOVA model should be given up and a multiple linear regression model with repeated measures on two factors: "situations" and "modes of response" should be used in the investigative program of interactional psychology. Being "nominal", these factors must be pre-coded (with the use of an "orthogonal coding") before they are introduced into the multiple linear regression model.

NOTES

1. Factor A includes the p-number of subjects. Each subject is treated as one level of the factor. Factor B includes the q-number of situations ($q = 11$), and factor C includes the r-number of the modes of response to situations ($r = 14$). The total number of combinations is $p \times 11 \times 14$, which equals $p \times 154$ of the combinations of the levels of factors.

2. The initial version of this test has been described by Tukey (1949).

3. Incidentally, we will not obtain explicit evaluations of variance components even in the case of the ANOVA design used by Endler et al. (1962) but considered within the framework of model *III*.

BIBLIOGRAPHY

1. Cattel, R.B., Scheier, I.H. *The meaning and measurement of neuroticism and anxiety.* New York: Ronald Press, 1961.
2. Cohen, J. Cohen, P. *Applied multiple regression/correlation analysis for the behavioral sciences.* Hillsdale, N.J.: L. Erlbaum, 1975.
3. Cornfield, J., Tukey, J.W. Average values of mean squares in factorials. *Annals of Mathematical Statistics* 1956, *27*, 907 — 949.
4. Cornbach, L.J., Gleser, G.C., Nanda, E., Rajaratnam, N. *The dependability of behavioral measurements.* New York: J. Wiley, 1972.
5. Ekehammar, B. Interactionism in personality from a historical perspective. *Psychological Bulletin*, 1974, *81*, 1026 — 1048.
6. Endler, N.S. Estimating variance components from mean squares for random and mixed effects analysis of variance models. *Perceptual and Motor Skills*, 1966, *22* 559 — 570.
7. Endler, N.S., Hunt, J. McV. Sources of behavioral variance as measured by the S-R Inventory of Anxiousness, *Psychological Bulletin*, 1966, *65*, 335 — 346.
8. Endler, N.S., Okada, M.A. Multidimensional measure of trait anxiety: The S-R Inventory of General Trait Anxiousness. *Journal of Consulting and Clinical Psychology*, 1975, *43*, 319 — 329.
9. Endler, N.S., Hunt, J. McV., Rosenstain, A.J. An S-R Inventory of Anxiousness. *Psychological Monographs*, 1962, *76* No. 17.
10. Ferguson, G.A. *Statistical analysis in psychology and education.* New York: McGraw-Hill, 1976.
11. Fleiss, J.L. Estimating the magnitude of experimental effects. *Psychological Bulletin*, 1969, *72*, 273 — 276.
12. Gaito, J. Expected mean squares in analysis of variance techniques. *Psychological Reports*, 1960, *7*, 3 — 10.
13. Gleser, G.C., Cronbach, L.J., Rajaratnam, N. Generalizability of scores influenced by multiple sources of variance. *Psychometrika*, 1965, *30*, 395 — 418.
14. Magnusson, D., Endler, N.S Interactional psychology: Present status and future prospects. In D. Magnusson, N.S. Endler (eds.), *Personality at the crossroads: Current issues in interactional psychology.* Hillsdale, N.J.: L. Erlbaum, 1977, 3 — 31.
15. Olweus, D. A critical analysis of the "modern" interactionist position. In D.

Magnusson, N.S. Endler (eds.), *Personality at the crossroads: Current issues in interactional psychology.* Hillsdale, N.J.: L. Erlbaum, 1977, 221 — 233.

16. Scheffé, H. *The analysis of variance.* New York: J. Wiley, 1959.
17. Spielberger, C.D. Anxiety: State-trait-process. In C.D. Spielberger, I.G. Sarason (eds.), *Stress and anxiety,* vol. 1. Washington: Hemisphere/Wiley, 1975a.
18. Spielberger, C.D. The measurement of state and trait axiety: Conceptual and methodological issues. In L. Levi (ed.), *Emotions - their parameters and measurement.* New York: Razaen Press, 1975b.
19. Spielberger, C.D., Gorsuch, R.L., Lushene, R.E. *Manual for the Stait-Trait Anxiety Inventory.* Palo Alto, Calif.: Consulting Psychologists Press, 1970.
20. Taylor, J.A. A personality scale of manifest anxiety. *Journal of Abnormal Psychology,* 1953, *48,* 285 — 290.
21. Tukey, J.W. One degree of freedom for nonadditivity. *Biometrica* 1949, *5* 232 — 242.
22. Vaughan, G.M., Corballis, M.C. Beyond tests of significance: estimatfng strength of effects in selected ANOVA designs. *Psychological Bulletin,* 1969, *72,* 204 — 213.
23. Winer, B.J., *Statistical principles in experimental design* (2nd. ed.). New York: McGraw-Hil, 1971.

PERSPEKTIVEN
DER PHILOSOPHIE

NEUES JAHRBUCH

Herausgegeben von
Rudolph Berlinger — Eugen Fink †
Friedrich Kaulbach — Wiebke Schrader
Johann-Heinrich Königshausen

USA/Canada: Humanities Press Inc., 171 First Avenue, Atlantic Highlands, N.J. 07716/USA

Japan: United Publishers Services Ltd., Kenkyu-sha Building, 9, Kanda Surugadai, 2-chome, Chiyoda-ku, Tokyo, Japan

BRD/Suisse/Austria: Verlag Königshausen & Neumann, Postfach 6007, D-8700 Würzburg, BRD

And others: Editions Rodopi B.V., Keizersgracht 302-304, 1016 EX Amsterdam, Telephone (020) — 22 75 07

Die Intention des Jahrbuches

Die Perspektiven der Philosophie eröffnen Forschern, die die Arbeit philosophischer Begründung und Rechtfertigung des Denkens auf sich nehmen, eine Publikationsmöglichkeit.
Das Jahrbuch begreift sich nicht als Schulorgan einer philosophischen Lehrmeinung. Kontroversen sollen ausgetragen werden, wenn sich dadurch Probleme konturieren und Argumente verschärfen. Das Jahrbuch sieht seine Aufgabe darin, an der Intensivierung eines wissenschaftlichen Philosophierens mitzuarbeiten.
Die Jahrbuchbände bringen jeweils Beiträge zu einem Rahmenthema und solche nach freier Wahl.
Rezensionen und Informationen zur philosophischen Diskussion sind vorgesehen.

Herausgeber und Redaktion

Das "Neue Jahrbuch" nimmt die Intentionen des ehemaligen Jahrbuchs "Philosophische Perspektiven" (1969-1973) auf.

INHALTSVERZEICHNIS

Dem Andenken von Edmund Husserl — Aron Gurwitsch — Wolfgang Cramer.

Eugen Fink (Freiburg i. Br.), Totenrede auf Edmund Husserl. Ludwig Landgrebe (Köln) und Jan Patočka (Prag), Edmund Husserl zum Gedächtnis. Zwei Reden. Fred Kersten (Wisconsin — Green Bay), Aron Gurwitsch. Ein philosophisches Portrait. Konrad Cramer (Heidelberg), Rede am Grabe meines Vaters. Hans Friedrich Fulda (Heidelberg), In memoriam Wolfgang Cramer.

Buchanzeigen und Diskussionen

NEUES JAHRBUCH Band 2 — 1976 . . . . Hfl. 80,—

INHALTSVERZEICHNIS

Ende oder Zukunft der Metaphysik

Franco Chiereghin (Padua), Ende oder Zukunft der Metaphysik. Wilhelm Ettelt (Freising), Die Rede vom metaphysischen Bedürfnis. Jacques D'Hondt (Poitiers), Wer läutet die Totenglocke? Dieter Lang (Würzburg), Zum erkenntnistheoretischen Problem der Metaphysik Schopenhauers. Martin Oesch (Hildesheim), Zur theologischen Philosophie. Josef Stallmach (Mainz), Zur Eindeutigkeit des Seinsbegriffes. Xavier Tilliette (Chantilly), Schelling und das Problem der Metaphysik.

Vermischte Abhandlungen

Gerhardt Funke (Mainz), Erscheinungswelt. Zur phänomenologischen Ästhetik, Teil II. Klaus Hartmann (Tübingen), Ideen zu einem neuen systematischen Verständnis der Hegelschen Rechtsphilosophie. Walter Hirsch (Kiel), Die Aktualität der klassischen Ästhetik. Paul Janssen (Köln), Zur Struktur geschichtsphilosophischer Aussagen. Thuri Lorenz (Würzburg), Das Bildnis des Sokrates. Magda Felice-Oschwald (Padua), Comte an d'Eichthal. Jean Servier (Montpellier), Das Drama des bürgerlichen Humanismus.

Dem Andenken von Heinz Heimsoeth und Eugen Fink

Wolfgang Janke (Köln), Vermächtnis Heinz Heimsoeth. Gerhart Schmidt (Bonn), Eugen Fink.

Buchanzeigen und Diskussionen

NEUES JAHRBUCH Band 9 - 1983 . . . . Hfl. 105,—

Dieser Band enthält Beiträge "Zur frühen Heidegger-Kritik", zur "Philosophie der Erziehung", es folgen im dritten Teil Abhandlungen zu vermischten Themen.

INHALTSVERZEICHNIS

Graduate Faculty Philosophy Journal

A JOURNAL OF
CONTINENTAL PHILOSOPHY

Werner Marx: Reflections on a
Non-Metaphysical Ethics

Michel Henry: Being as Production

Alaine Touraine: Social Movements,
Revolution, Democracy

Subscriptions for individuals: $9.50 per year
Libraries and institutions: $12.50
Single copies of back issues: $5.00
Address all correspondence to: The Editor,
Graduate Faculty Philosophy Journal, 65 Fifth Avenue,
N.Y., N.Y. 10003

GRAZER
PHILOSOPHISCHE STUDIEN

BAND 23   1985   VOLUME 23

Herausgeber   *Editor*

Prof. Dr. Rudolf HALLER, Institut für Philosophie, Universität Graz, Heinrichstraße 26,
A-8010 Graz, Österreich/Austria.

PRAGMATICS
Handbook of Pragmatic Thought
Edited by Herbert Stachowiak
(All Articles in German or English)

The "pragmatic turning point" in philosophy in the second half of the twentieth century is a testimony to the far-reaching changes in the leading objectives of philosophical reflection. The handbook PRAGMATICS in several volumes takes the effort to illuminate the basic idea of pragmatic thinking and to prove it scientifically, thereby commending it as *the* guiding principle in the construction of a model designed to cope with the problems of survival ahead from the perspective of rationally reflected, practical experience. The handbook PRAGMATICS sets out to achieve this objective in three ways: 1. by offering an historically systematic presentation of the development of pragmatic thought as well as the differentiation of the pragmatic approach over the past three decades; 2. by comprehensively extending the field of enquiry to encompass the whole area of operative, constructive and purposeful thought and action upon which human decisions are made; 3. through the systematic registering, evaluation and further application of present pragmatic tendencies in the fields of philosophy of the individual sciences and in social and political practice.

Leading representatives of various scientific disciplines from twelve different countries, in particular from the USA, have participated in this standard reference work. All the articles are appearing for the first time in German or English. The uniform treatment of the structure, recommended reading list and graphs are a further aid to the user. So, too, is the register of authors, names and subjects at the end of each volume.

THE INDIVIDUAL VOLUMES

Volume I: **Pragmatic Thought from its Origins to the Eighteenth Century**
Edited by Herbert Stachowiak in cooperation with Claus Baldus. Introduction by Herbert Stachowiak. 1986. 568 pages. Bound. Subscription price for subscribers to the complete works DM 158,–. Single volume price DM 178,–.

Part 1: *Cultures and Religions*. Articles by W. Dupré, J. Nagl, E. Hornung, W. Stark. Part 2: *Classical and Early Christian Thought*. Articles by E. Lang, E. Martens, K. Mainzer, A. Baruzzi, F. Körner. Part 3: *The Middle Ages: Transition and Awakening*. Articles by K. Bosl, G. Endreß, . Burrel, E. and W. Leinfellner, U. Köpf, N. Herold. Part 4: *Renaissance, Reformation and Enlightenment*. Articles by R. zur Lippe, R. Stupperich, R. K. DeKosky, H. Sachsse, J. Mittelstraß and P. Schroeder-Heister, C. Axes, S. Müller. Part 5: *Transcendental Idealism*. Articles by Kaulbach, C. Baldus, C. Cesa.

Volume II: **The Rise of Pragmatic Thought in the Nineteenth and Twentieth Century**
Edited by Herbert Stachowiak in cooperation with Claus Baldus. Introduction by Herbert Stachowiak. 1986. Approx. XVI, 464 pages. Bound. Subscription price for subscribers to the complete works approx. DM 158,–. Single volume price approx. DM 178,–.

Part 6: *Pragmatic Emancipation*. Articles by K. and D. Claessens, H. Ebeling, R. E. Butts, L. Schäfer, C. Eisele. Part 7: *The Rise of Pragmatism*. Articles by R. Almeder, D. Sidorsky, E. Rochberg-Halton, M. Nadin, I. L. Horowitz, C. F. Gethmann, P. Janich, H. Lenk and M. Maring. Part 8: *Modern Systematisings*. Articles by G. Meggle, R. M. Martin, A. Rapoport, T. Pszczolowski, H. Stachowiak.

Volume III: **General Philosophical Pragmatics**
Edited and introduced by Herbert Stachowiak
Part 1: *Forms of Reflection and the Problem of Rationality*. Articles by C. F. Gethmann, J. Habermas. Part 2: *Practical Philosophy*. Articles by F. Stoutland, A. Pieper, H. Stachowiak, W. Schmied-Kowarzik. Part 3: *Man as Sufferer, Seeker of Meaning and Optimist*. Articles by W. C. Zimmerli, K. Wuchterl, E. A. Moutsopoulos, A. Schwan. Part 4: *Philosophy of History, Society and Law*. Articles by K. Acham, H. Stachowiak, R. S. Summers. Part 5: *Cosmology, Ecology and the Philosophy of Technique*. Articles by J. Audretsch, D. Birnbacher, H. Sachsse. Part 6: *The Paradigm of Recent Pragmatism*. Articles by H. S. Thayer, H. Lenk.

Volume IV: **Philosophy of Language, Linguistic Pragmatics and Formative Pragmatics**
Edited and introduced by Herbert Stachowiak
Part 1: *Pragmatic Linguistic Philosophy*. Articles by H. J. Schneider, K. O. Apel, D. Wandschneider, H. N. Castañeda. Part 2: *Linguistic Pragmatics*. Articles by H. Bickes, W. Vossenkuhl, W. Kummer, J. S. Petöfi, A. Sakaguchi. Part 3: *Semiotics and logical Pragmatics*. Articles by M. Nadin, R. M. Martin, C. E. Alchourrón u. E. Bulygin, C. F. Gethmann, N. Tennant and P. Schroeder-Heister, R. Hilpinen. Part 4: *Mathematical Pragmatics*. Articles by R. Inhetveen, C. Thiel, V. Pittioni.

Volume V: **Pragmatic Tendencies in Scientific Theory**
Edited and introduced by Herbert Stachowiak
Planned are Parts 1: *The Dynamics of Science: Pragmatic Reflections; 2: Recent Pragmatic Concepts of Epistemology; 3: Old and New Problems in the Pragmatic Paradigm*.

Please order special prospectus!

As of 1.3.1986

FELIX MEINER VERLAG · 2000 HAMBURG 76

THE JOURNAL OF THE BRITISH SOCIETY FOR PHENOMENOLOGY

An International Review of Philosophy and the Human Sciences

EDITOR: WOLFE MAYS

Volume 17 Number 1, January 1986
Karl Jaspers: Philosophy and Psychiatry

Articles

The Central Gesture in Jaspers' Philosophy, by Jeanne Hersch
The Significance of Jaspers' Philosophy for our Time, by Gerhard Huber
Karl Jaspers and the Quest for Philosophic Truth, by Alfons Grieder
Karl Jaspers' Influence on Psychiatry, by Wolfram Schmitt
The Negative Effects on Psychiatry of Karl Jaspers' Development of *Verstehen*, by
    F. A. Jenner, A. C. Monteiro and D. Vlissides
Jaspers and Nietzsche, by M. Skelton-Robinson

Reviews and Notes

The *JBSP* publishes papers on phenomenology and existential philosophy as well as
contributions from other fields of philosophy. Papers from workers in the Humanities and human sciences interested in the philosophy of their subject will be welcome.
All papers and books for review to be sent to the Editor: Dr. Wolfe Mays, Department
of Philosophy, University of Manchester, Manchester M13 9PL, England. Subscription and advertisement enquiries to be sent to the publishers: Haigh and Hochland
Ltd., The Precinct Centre, Oxford Road, Manchester 13, England.

ALOIS DEMPF
Metaphysik
Versuch einer problemgeschichtlichen Synthese
Amsterdam/Würzburg 1986. 332 pp. (Elementa Band 38). ISBN: 90-6203-702-6   Hfl. 70,—

In Zusammenarbeit mit Christa Dempf-Dulckeit.
Die vorliegende Arbeit ist die letzte des 1982 verstorbenen Münchner Philosophie Historikers Alois Dempf. Wie der Untertitel ausweist, handelt es sich um eine umfassende Synthese historisch-systematischer Art. Es geht dem Autor nicht darum, neue einzelwissenschaftliche Erkenntnisse vorzulegen, sondern Zusammenhänge und Abhängigkeiten unseres Denkens und unseres heutigen Wissenschaftsbildes so in den Blick zu rücken, daß eine Entwicklung durchsichtig wird, die dem Leser eine Standortsbestimmung ermöglicht. Auf eine Einleitung, die — soweit das überhaupt möglich ist — in einfacher Art philosophische Grundpositionen zu klären versucht, folgen drei Kapitel, die die großen, perennen Themen der Metaphysik entwickeln. Welt- und Lebenslehren werden problemgeschichtlich behandelt und schließlich wird abschließend eine allseitige Anthropologie in ihrer Verflechtung mit der Theologie entworfen. Hierbei steht eine mehrdimensionale Menschenstruktur im Mittelpunkt, die Voraussetzung ist für die Bewältigung der Probleme in einer Zeit der mechanisierten Steuerung aller Umwelten. Das Buch soll zu dem Versuch beitragen den beklemmenden l'art pour l'art — Standpunkt modernen philosophischen Denkens zu überwinden und auf kritisch-realistischer Grundlage die lebendigen Wurzeln unserer geistigen Existenz freizulegen.

Cornelia Eva KEIJSPER
Information Structure
With examples from Russian, English and Dutch
Amsterdam 1985. 385 pp. (Studies in Slavic and General Linguistics Vol. 4).
ISBN: 90-6203-627-9   Hfl. 90,—

Keizersgracht 302-304
1016 EX AMSTERDAM-HOLLAND

RUDOLF LÖBL
Die Relation in der Philosophie der Stoiker
Amsterdam/Würzburg 1986. 150 pp. (Elementa Bd. 37). ISBN: 90-6203-587-6
Hfl. 35,—

Denken in Beziehungen spielt in Mathematik und Logik eine herausragende Rolle und hat in den exakten Naturwissenschaften immer mehr an Bedeutung gewonnen. Nur in der Philosophie hat die Relation noch nicht das Gewicht, das ihr — auch im Blick auf die Wissenschaften — eigentlich zukommen sollte. Relationstheoretisches Denken erweist sich immer mehr als ein fruchtbarer Weg des Philosophierens, selbst wenn man nicht mit J.J. Schaaf die Relation als den entscheidenden Grundbegriff der Philosophie betrachten wollte. Diese Art zu denken bewährt sich nicht nur in der Bewältigung gegenwärtiger philosophischer Probleme, sondern auch in der Interpretation vergangenen philosophischen Gedankenguts. Einer relationstheoretischen Interpretation der stoischen Philosophie erweist sich, daß Beziehung und Bezogensein nicht nur eine prinzipielle innere Struktur dieser Philosophie ist, sondern auch das einigende Band darstellt, das die verschiedenen Ausformungen, die diese Philosophie in ihren Vertretern erfahren hat, ebensowohl von innen her verbindet als es auch ihre drei Teile- Physik, Logik, Ethik- untereinander von innen her zusammenhält. Kernbegriffe wie Physis, Logos, Heimarmene und Sympatheia (um nur diese vier zu nennen) rücken dabei in ein neues Licht, das die inneren Strukturen und äußeren Konturen dieser geistigen Bewegung deutlicher hervortreten läßt, die ein halbes Jahrtausend lang in der Antike lebendig war und darüber hinaus anregend und fruchtbar im europäischen Denken weitergewirkt hat.

HEINRICH MIDDENDORF
Phänomenologie der Hoffnung
Vorwort Paul Janssen
Amsterdam/Würzburg 1985. 99 pp. (Elementa Band 40). ISBN: 90-6203-617-1
Hfl. 25,—

Keizersgracht 302-304
1016 EX AMSTERDAM-HOLLAND

ON SHMUEL HUGO BERGMAN'S PHILOSOPHY
edited by A. Zvie Bar-On
Amsterdam 1985. 242 pp. ISBN: 90-6203-947-2   Hfl. 40,—

Contents: A. Zvie Bar-On: A Bio-Bibliographical Note. Nathan Rotenstreich: Between Construction and Evidence. Rudolf Haller: The Philosophy of Hugo Bergman and the Brentano School. A. Zvie Bar-On: From Prague to Jerusalem. Joseph Agassi: On Hugo Bergman's Contribution to Epistemology. Yirmiyahu Yovel: Reason as Necessary and Insufficient. Gershon Weiler: Bergman as a Historian of Philosophy. Joseph Horovitz: A Criticism of Shmuel Hugo Bergman's Account of Nicolaus Cusanus. Ze'ev Levy: S.H. Bergman on the Relation between Philosophy and Religion.

HENDRIK KAPTEIN
Ethiek tussen twijfel en theorie
Amsterdam 1985. 192 pp. ISBN: 90-6203-578-7

Hfl. 29,50

Denken over ethiek komt voort uit onzekerheid over wat we moeten doen. Normatieve ethiek kan voorlopig worden bepaald als bestaande uit die regels en beginselen die voorrang hebben boven alle andere regels, doelen etc.. Bepalingen van ethiek in termen van medemenselijkheid zijn circulair. Metaethiek is logische analyse van morele uitdrukkingen, maar ook methodologie, die gaat over manieren waarop normatieve ethiek zou kunnen worden gerechtvaardigd. Descriptieve ethiek is geen normatieve ethiek. We hebben al allerlei, bijzondere en algemene, bewuste en onbewuste, morele gedachten en gevoelens. Feiten zorgen in hoofdzaak voor dat logische verband, waardoor een zo goed mogelijke ontwikkeling en uitdrukking van morele gedachten en gevoelens. Feiten zorgen in hoofdzaak in voor dat logische verband, waardoor een morele theorie kan ontstaan. Een reflectief equilibrium is gedachten en gevoelens. Het keurslijf van zo'n theorie heeft een regimenterende en objectiverende werking. Objectiviteit is niet hetzelfde als betrokkenheid op objecten. Zolang volledige morele theorieën door gebrek aan feitenkennis niet beschikbaar zijn, kunnen we ook niet weten of die objectief of subjectief, relatief zijn. Wel zijn er elementaire morele uitgangspunten waar niemand echt aan twijfelt. Morele skepsis is vaak nogal academisch. Heel algemene vragen als die naar de aard van kennis methodologie, gaat ook over zichzelf. Zij laat zien, dat maatstaven van rationaliteit in normatieve ethiek voor een belangrijk deel overeenkomen met maatstaven van rationaliteit en kennis in het algemeen.

Keizersgracht 302-304
1016 EX AMSTERDAM-HOLLAND

INTERNATIONAL BIBLIOGRAPHY OF AUSTRIAN PHILOSOPHY
IBÖP 1974 – 1975
INTERNATIONALE BIBLIOGRAPHIE ZUR ÖSTERREICHISCHEN
PHILOSOPHIE
edited by
Wolfgang L. Gombocz, Rudolf Haller, Norbert Henrichs
Amsterdam 1986. 172 pp. (Studien zur Österreichische Philosophie
Supplement 1). ISBN: 90-6203-747-X Hfl. 70,—

Contents: Teil 1: G.L. Gombocz, R. Haller, N. Henrichs: Vorwort.
Literaturhinweise zur IBÖP / *Bibliographical Notes*. Bildnis / *Frontispiece*:
Bernard Bolzano (1781-1848). Jan Berg und Edgard Morscher: Aufsatz /
Article: Bernard Bolzano — Der österreichische Philosoph. Hinweise für
den Gebrauch der Bibliographie und Register. *How to Use Bibliography and
Index*. Teil 2: Bibliographie / *Bibliography*. Dokumente 1-999 / *Items No. 1-
999*. Sachregister / *Subject Index*. Namenregister / *Index of Names*.

CHRISTIAN VON EHRENFELS
Leben und Werk
herausgegeben von Reinhard Fabian
Amsterdam 1986. 286 pp. (Studien zur österreichischen Philosophie Band
8). ISBN: 90-6203-856-5 Hfl. 60,—

Inhaltverzeichnis: Vorwort. Reinhard Fabian: Leben und Wirken von
Christian v. Ehrenfels. Ein Beitrag zur intellektuellen Biographie. Theo
Herrmann: Die Gestalttheorie von Christian v. Ehrenfels im Lichte
moderner Kognitionspsychologie. Kevin Mulligan und Barry Smith: Mach
und Ehrenfels: Über Gestaltqualitäten und das Problem der Abhängigkeit.
Peter M. Simons: Mathematik als Wissenschaft der Gestalten. Roderick M.
Chisholm: Reflections on Ehrenfels. Unity of Consciousness. Barry Smith:
The Theory of Value of Christian von Ehrenfels. Rudolf Haller: Zu
Ehrenfels' Ästhetik. Gerhard J. Winkler: Christian von Ehrenfels als
Wagnerianer. Reinhild Rug und Kevin Mulligan: Theorie und Trieb —
Bemerkungen zu Ehrenfels. J.C. Nyíri: Geschichtstypologische Bemerkun-
gen zur böhmischen Frage.

Keizersgracht 302-304
1016 EX AMSTERDAM-HOLLAND

M. GLOUBERMAN
Descartes: The Probable and the Certain

Amsterdam/Würzburg 1986. 374 pp. (Elementa Band 41). ISBN: 90-6203-700-X   Hfl. 80,—

This book comprises an integrated interpretation of the core of Descartes' philosophy: its epistemology, ontology, and semantics. Pivotal is an unorthodox construal of the Cartesian contrast between certainty and uncertainty. It is maintained that the transition from uncertainty or 'probability' to certainty involves an alteration in categorial structure, and an attendant ontological shift. In the course of the journey, a roughly Platonic system of categories supplants a roughly Aristotelian one, and a non-realist ontology gives way to a realist ontology. To back up this novel reading, the thematic basis and methodology of the influential 'analytic' approach to Descartes are criticised. Specifically, the usual sceptical construal of Cartesian doubt is undermined; and the utility of the entrenched empiricist/rationalist typological duality challenged. More generally, it is shown that an analytic approach presupposes substantive philosophical theses which prejudice the treatment of pre-Kantian texts. The Cartesian results make possible an evolutionary account of the lines of development from pre-Kantian thought to Kant's 'critical' ensemble. In a fashion which removes much of the puzzlement surrounding the provenance of Kant's mature view, the elements thereof are shown to derive fairly directly from the left side of the Cartesian probable/certain contrast.

K.R. SCHMITT
Death and After-Life in the Theologies of Karl Barth and John Hick
A Comparative Study

Amsterdam 1985. 230 pp. (Amsterdam Studies in Theology. Volume 5). ISBN: 90-6203-528-0   Hfl. 50,—

Keizersgracht 302-304
1016 EX AMSTERDAM-HOLLAND

Erkenntnis

An International Journal of Analytic Philosophy

Editors   ISSN 0165–0106
CARL G. HEMPEL, WOLFGANG STEGMÜLLER, and WILHELM K. ESSLER

Erkenntnis is a philosophical journal publishing papers on foundational studies and scientific methodology covering the following areas:
— that field of philosophy associated today with the notions of 'Philosophy of Science' and 'Analytic Philosophy' (in a wide sense);
— the philosophy of language, of logic, and of mathematics;
— the foundational problems of physics and of other natural sciences;
— the foundations of normative disciplines such as ethics, philosophy of law and of aesthetics;
— the methodology of the social sciences and the humanities;
— the history of scientific method.

Subscription Information
1984, Volume 21 (6 issues)
Institutional rate: Dfl. 190,— / US $ 76.00 including postage and handling
Private rate: Dfl. 65,— / US $ 26.00 including postage and handling
Private subscriptions should be sent direct to the publishers.

 # D. Reidel Publishing Company

**P.O. Box 17, 3300 AA Dordrecht, The Netherlands
190 Old Derby St., Hingham, MA 02043, U.S.A.**

STUDIA LOGICA

An international quarterly for symbolic logic

Editors: Dagfinn FØLLESDAL, Helena RASIOWA, Krister SEGERBERG, Ityszard WÓJCICKI (Editor-in-Chief), Grzegorz MALINOWSKI (Executive Editor)

Volume 43, Numbers 1/2 (1984)

Contents

Published by Ossolineum Publishing House and D. Reidel Publishing Company. Subscription orders: from the socialist countries — RSW "Prasa-Książka-Ruch", Wronia 23, 00-840 Warszawa, Poland; from outside the socialist countries — D. Reidel Publishing Company c/o Kluwer Academic Publishers Group, Distribution Centre, P. O. Box 322, 3300 AA Dordrecht, Holland.

Printed in the United States
By Bookmasters